"It has reframed my thinking around what it means to scale sustainability globally, and I share Kathryn's esteem for the common good of the planet as a lens for rewiring our systems and in turn, our behaviors, to achieve universal impact."

Lindsey Peterson, *Global President of Earthforce at Salesforce*

"Kathryn ably stares down the 'hard problem' of sustainability... bringing her pragmatic systems thinking to the challenge. The five transformational forces will re-shape the questions you ask in your business, in policy making, decision making and strategy setting. How might we harness these transformational forces to create a sustainable planet?"

Meredith England, *Director of Systems Innovation, Climate-KIC Australia*

"Kathryn has done an amazing job of not only sounding the climate alarm bells but she also intelligently processes how digital technologies and transformations are crucial solutions to scale the change needed."

Eddie Listorti, *CEO, Viridios Capital Global*

I0130385

Digitalizing Sustainability

Digitalizing Sustainability outlines why 'business as usual' isn't working and sets out five Transformational Forces which can be used to innovate and scale sustainability solutions using digital means.

This transformation will be powered by a range of digital technologies that have the potential to ideate, propel and scale sustainability solutions in an exponential manner over the next decade. This book introduces the Five Forces of Digital Transformation. These forces all share a common root – they are powered by digital technologies that enable them to operate at the speed and scale that we need to achieve global transformation for people and planet. Working together, these forces help overcome many of the barriers and pitfalls that humanity has faced in trying to achieve sustainability 2.0 and will help you tackle these questions:

1. *Why has sustainability 1.0 failed us?*
2. *What are the bugs and bad code in our operating systems that we must address to have any hope of creating a sustainable civilization?*
3. *What are the Five Forces of Digital Transformation that will forge a sustainable and resilient future for our people and planet?*
4. *What are the core priorities going forward for coalitions of the willing to harness these forces for sustainability and resilience?*

Disruptive and innovative, this book provides readers with a clear path forward to a sustainable digital future. It will be of great interest to anyone interested in the adoption of digital technologies for driving solutions to global sustainability challenges.

Kathryn Sforcina began her journey with radical transformation as a serial entrepreneur and has 20+ years experience as a c-level executive and board member. Kathryn is presently the Founding Director of Transforming Tribes (*transformingtribes.com*), a management consulting firm that transforms the way modern-day tribes relate to people, purpose, place and planet by exploring the intersection of ancient wisdom, modern science, human connection and emergent technologies.

Digitalizing Sustainability

The Five Forces of Digital
Transformation

Kathryn Sforcina

Routledge
Taylor & Francis Group
LONDON AND NEW YORK

earthscan
from Routledge

Cover image: © Shutterstock

First published 2023
by Routledge
4 Park Square, Milton Park, Abingdon, Oxon OX14 4RN

and by Routledge
605 Third Avenue, New York, NY 10158

Routledge is an imprint of the Taylor & Francis Group, an informa business

British Library Cataloguing-in-Publication Data
A catalogue record for this book is available from the British Library

ISBN: 978-1-032-03482-9 (hbk)
ISBN: 978-1-032-03483-6 (pbk)
ISBN: 978-1-003-18752-3 (ebk)

DOI: 10.4324/9781003187523

Typeset in Times New Roman
by Newgen Publishing UK

For my children Jesika, Lyvia and Moses. Thank you for your love and for the privilege of being your mother. You, your generation and all life yet to come into being, are the inspiration that makes me want to leave the world a better place than I found it.

For Pachamama. Like every other living thing in your world, I owe you my life. Thank you for how you sustain us, nurture us and hold us. The way you keep giving back to us – no matter the degradation and challenge we bring to your door – is a daily lesson in love and grace that I am deeply grateful for the privilege to witness and learn from. Thank you. May we continue to learn from you, grow with you and walk more gently in our relationship with you. May we learn how to use our advancement as a species to protect you, care for you and truly step into our role as stewards of this planet. For your sake, and for our own. Aho.

Contents

x *Contents*

Acknowledgments

First, I wish to take a moment to acknowledge the Gadigal people of the Eora nation, whose country is the place upon which most of this book was written. I acknowledge they never ceded sovereignty. I recognize their continuing connection to the land and waters, and thank them for protecting this coastline, bushland and its ecosystems since time immemorial. I pay my respects to Elders past, present and emerging and extend that respect to all First Nations people present around the globe today.

I also want to acknowledge that it has literally taken a village to bring this book into the world and as such, would like to sincerely thank my publishers, Taylor and Francis Group, and the team lead by Annabelle and Jyotsna that have helped make this book possible. Your support and encouragement have gone above and beyond. Thank you so much for believing in me and helping me bring this project to life.

To my dear friend David whose countless hours of educated debate and willingness to be my most constant (yet kind and fair) critic, enabled me to explore all the corners of my brain and find my courage to express it on paper. Your friendship and support throughout the months of COVID lockdown and your (extraordinary) generosity with your time spent helping me unpack the concepts that now fill these pages, is a huge component of what made it possible for me to bring completion to this body of work. For that, I am eternally grateful.

To my beautiful son Moses, you have always been my number one fan. Your support, love and belief in me have made me the human I am today. Thank you for all the hours of family time you've sacrificed over the past couple of years to allow me the time to think and write. I love you so much xx. To my amazing daughters Lyvia and Jesika for cheering me on from the sidelines and inspiring me with the way you show up in the world. You are the kind of young women the world needs more of. No matter how far away you both are, you have my heart, always xx.

To the dearly loved members of my soul tribe, Carly, Josh, Jordie, Meredith, Rebecca, Pablo, Michael C., Jodie, and Becky who picked me up and brushed me off whenever the "self-doubt gremlins" came to visit and for all the times you reminded me that this message was important and needed to be heard. Thank you, thank you, thank you.

To Bechin, Ricky, Riv, Sal, Kate, Sam and Steve. Thank you for holding space for me as I explored the deepest places within myself. Thank you for your mentorship, guidance, acceptance, wisdom and love. Thank you for enabling the landscapes to explore paradigm shifts in beliefs, perspective and more, to exist in me. Thank you for helping me find my path. Thank you for helping me find my voice. And most of all, thank you for your dedication to your own journeys and the beautiful gift that is the medicine you bring into the world.

Finally, I also wish to acknowledge the Stalking Wolf lineage and that of the Jaguar lineage whose ancient ways and wild wisdom have taught me so much about what it means to belong to country and how critical to my well-being, and that of all humanity, having a relationship with the more-than-human is. Thank you deeply for the gift of your medicine also. Your ways have brought me into so many new understandings about life, the connection to all and ignited the spark in me to share my own medicine with the world. This book is as much for you as it is because of you.

Part One

Understanding the 'Hard Problem' for Sustainability

1 Tracking the trajectory of Spaceship Earth

The past 50 years in review

Since 1972, world leaders have signed some 500 internationally recognized environmental conventions. These agreements include 61 related to the atmosphere; 155 to biodiversity; 179 to chemicals, hazardous substances and waste; 46 land agreements; and 196 conventions related to water. After trade agreements, the environment has become the most prolific area of global rule-making that this century has seen. [1]

At the beginning of 2020, the scientific community adopted a new mantra – moving from norms and agreements to action. They dubbed the next ten years "the decade of action" – ten years to protect the planet while advancing global prosperity. Whether this is slowing climate change, protecting biodiversity, reducing pollution, or achieving Sustainable Development Goals (SDGs), ten years was the timeframe agreed upon to achieve a global paradigm shift away from a dominant capitalist mindset, and its requisite of infinite growth on a finite planet.

It feels appropriate then to begin the opening pages of this book by putting these 500+ agreements, and two years of progress, toward the "decade of action" under review.

In doing so, I find myself asking one simple question:

How are we tracking so far?

Have we been able to "bend the curve" downwards on greenhouse gas emissions, deforestation, land degradation, habitat loss, or the myriad of other environmental variables of concern? Are we transforming economic incentives and human behaviors at the speed and scale needed to meet global social justice, climate, biodiversity and pollution targets by 2030?

In other words, as we orbit the sun and fly across the galaxy in our "Spaceship Earth", have we, the captains of the spaceship, effectively protected the natural life support systems required to sustain our civilization?

Unfortunately, the answer is "no". [2] Not even close.

DOI: 10.4324/9781003187523-2

As this book goes to print, the true face of climate change is beginning to show its ugly self. Uncontrollable wildfires leave a trail of destruction in many parts of the world and increase in severity year-on-year. We've all watched in disbelief as climate change has disturbed the directional flow of jet streams, ancient systems that have existed since time immemorial, in the process causing a deadly heat dome to encapsulate the US and Canada. We've also witnessed the devastating impact of flooding and drought across multiple countries including those in Europe and Australia.

The hottest seven years recorded in our planet's known history have also occurred in the seven consecutive years since 2014. [3] Every year is now noticeably warmer than the last. And as the consequences of our human interference continue to play out, we will only see the planet respond with increasing levels of unpredictability and extreme events.

Indeed, as Andrew Zolli, Chief Impact Officer at the satellite company, Planet, likes to remind us, in almost every keynote address he has given in the past couple of years, "*We now live in an age of spikes*". By "spikes", Zolli is referring to the alarming level of exponential environmental changes, and their consequences, that are now occurring around us, on an almost daily basis.

The reason for so much alarm when dealing with exponential change is very simple; when the rate of change is exponential, the impact of this change doubles over certain intervals of time. This is known as the "doubling time".

For example, during the first peak of the COVID-19 pandemic, the doubling time in the US was 2.68 days, prior to widespread mitigation efforts. Yet, even at this rate, most of us dramatically underestimated how long it would take the disease to spread from one person to the entire human population.

In my own living room experiment, intuitive guesses from my family members and educated colleagues ranged from 2–10 years. The actual answer was approximately 80 days. There was such a large margin of error when making our predictions because the cognitive capabilities of our minds simply don't process and calculate exponential rates of change. They are instead hard-wired to think in linear terms of progression.

This is why anything that has hit an exponential trajectory, in the context of humanity and our existence on Spaceship Earth, is so powerful. It's not because it's necessarily fast. But we, as humans, cannot fathom the speed and scale of its impact, and, as a result, do not respond with the proportionate level of action required. By the time we realize that 'something needs to be done', it is usually too late to implement any form of action that could easily make an impact. So while the COVID-19 pandemic propelled the concept of exponential change into the public limelight, exponential change itself has been a constant background theme throughout modern history.

The power of human influence on our planetary life support systems

For quite some time, most of humanity has been on an upward, exponential curve of growth with respect to population size, consumption of goods, and extraction of natural resources. Our influence has impacted everything from the makeup of ecosystems to the geochemistry of Earth, from the atmosphere, to the land, and even the ocean. Our collective impact is now so large that we've become our own geological force. Scientists have even coined a new term to define this time in the planet's history. We now live in what is known as the "Age of the Anthropocene"; a label now widely used to describe the exponential scale of our human influence on the planet. [4]

These exponential growth curves that our human influence have instigated can be largely attributed to the technological advancements that we've made. Collectively, they are known as the "Great Acceleration". [5] Yes, we are the most advanced version of our species ever to have existed. We can instantaneously communicate with anyone on the globe, track our movements over the surface of the Earth in real time, genetically engineer living things, build robots that build robots, get from A to B in driverless vehicles, and send drones into cloud formations to produce our own rain cycles. It is also highly likely that in less than ten years' time a single computer will have the same processing capacity as the brain power of the entire human race combined. The technological and intellectual achievements of the human race are commendable, and quite clearly far exceed any of the planet's other residents.

Conversely, it can be argued that all technology seems to have given us so far is more power to exploit Earth's natural resources, in order to meet our exponentially growing food, energy and consumer needs. Right now, if we were to look at a planetary-scale dashboard of Spaceship Earth's life support systems, it would reveal a series of exponential impact curves of which we should be far less proud. Nearly every single warning indicator is now flashing red. Indeed, the true cost of our human advancement has been the exponential degradation of our planet. [6]

But let's return to the topic of climate change. Our planet is still heading for a temperature rise in excess of 3°C or more this century – far beyond the Paris Agreement goals of limiting global warming to well below 2°C, and pursuing 1.5°C.

In the domain of biodiversity, nature and ecological health, human activity has resulted in the severe alteration of more than 75 percent of Earth's land areas. The biomass of wild mammals has fallen by 82 percent, natural ecosystems have lost about half their area, 90 percent of our fish-stocks have been fished from our oceans, and a million species are at risk of extinction. [7] And if this isn't enough, the World Wildlife Fund (WWF) has now declared that humanity is causing the sixth mass extinction of life on Earth, [8] while its 2014 Living Planet Report noted that population sizes of vertebrate species have declined by 52 percent over the last 40 years. [9]

Meanwhile, global chemical production capacity almost doubled between 2000 and 2017. [10] Our industrial farming methods, such as tilling and fertilizer use, will have removed all farmable topsoil within the next 60 years. [11] These same fertilizers are also entering coastal ecosystems, and have produced more than 400 ocean 'dead zones', totaling over 245,000 km^2 – a combined area greater than that of the United Kingdom. Marine plastics pollution has increased tenfold since 1980; by 2050, the oceans could well have more plastic than fish. [12]

Of 45 megacities monitored, only four achieved standards in line with World Health Organization (WHO) guidelines for acceptable air quality. The term "Urban Heat Island Effect" is also a major consideration. As temperatures rise, significant infrastructure, such as transport and energy networks, coupled with the endless concrete jungles also known as the "suburban sprawl", magnifies these temperatures, causing major structural damage to buildings, buckled roads and train lines, accelerated evaporation of vital water sources, and power grids to falter. Even the loss of life for the more vulnerable people that live in these areas can be attributed to it. [13]

As more time elapses, the accumulating pollution from chemicals and waste, as well as the changing climate and biodiversity loss, are becoming increasingly interrelated. As these elements become entwined, they are reinforcing each other through complex feedback loops, which in turn continue to drive an ever-accelerating rate of exponential change.

Yet the problems resulting from our influence don't end there.

The climate crisis also brings with it significant challenges of a humanitarian nature. "Climate Refugee" is the term being used to describe the estimated 1.2 billion people who are predicted to be displaced by 2050 due to climate change. [14]

Social justice issues, and how they both impact, and are impacted by, climate change, are also of major concern. In our current economic model, wealth inequality and the gap between the developed world and the developing world will only continue to widen, as will class structures within urbanized areas. Resources will continue to become more and more scarce, and prices will continue to climb based on this leaner supply and greater demand – it's just how our capitalist economy is wired. This leaves very little opportunity for the impoverished or underprivileged.

All of these convergent pressure points simultaneously rising to unmanageable levels will cost our global economy trillions of dollars, [15] and could generate multiple sources of untold social conflict and violence. [16]

Just as none of our systems can function independently from one another, nor do the impacts from the problems attributed to these systems act in isolation. Separateness in our natural and human world simply does not exist. Considering this component of our ecosystem, each seemingly separate crisis we face will only continue to have an ongoing, and worsening, cascade effect on other aspects of the delicately balanced ecology of the planet. This puts into context the one big problem that we all now face:

> Human influence has placed our species and our host planet on two equal and yet opposing exponential trajectories – one of exponential growth in the name of human advancement, and one of exponential degradation to the detriment of our planet. Neither of these can be sustained if we wish to thrive, or even survive, on our planetary spaceship.

Why has Sustainability 1.0 failed?

As I ponder the precarious position that our impact on Earth has placed on our species, I wonder; despite nearly 50 years of concerted effort, why haven't we been capable of making the positive exponential impact that is so demonstrably needed?

I believe the single largest problem our sustainability efforts have come up against is an inability to embrace systems thinking, [17] and catalyze systems level transformations – basically, the 'hard problem' of the sustainability field.

Instead, practitioners have largely focused on the 'easier problem' of sustainability, creating solutions that were hyper-focused on addressing sustainable products and services, without necessarily triggering systems level transformation.

For example, solar panels, electric cars, and LED lights are all fantastic innovations, but they don't necessarily lead to the transformation of underlying systems, incentives and behaviors. This tunnel-visioned approach to sustainability didn't consider how to enable solutions to scale or interoperate across other systems and solutions; effectively preventing the widespread adoption of such innovations as could have led to exponential change.

Those efforts that did attempt to catalyze systemic change often began with the idea that sustainability could be achieved simply by 'greening' our existing economic systems, or by 'greening growth'. This perspective assumes that we could take exactly the same economic incentives, entrenched interests, and assumptions of infinite growth, and use government policies, regulations and taxes to achieve global sustainability outcomes.

However, there is an immediate problem with this approach; it fails to address the underlying systemic drivers and feedback loops that fundamentally motivate and shape human values, norms, behaviors, habits and cooperation.

A further limitation born from this inability to embrace systems thinking is that Sustainability 1.0 often assumed that natural resources and ecosystems could be managed as complicated systems in a 'steady state' of predictable inputs and outputs. This is misguided, as it fails to account for the reality of how ecosystems really function. Sustainability 1.0 thus largely failed to embrace ecosystems as dynamic systems that change over time and space. Sustainability 1.0 also struggled to account for the many complex interacting components of natural systems, which feature often unpredictable interrelationships and interdependencies that are fundamental to their operation and outcomes.

Indeed, Sustainability 1.0, as a singular organizing concept to sustain the life support system of our Spaceship Earth, was simply immature. In order to work, it needed to be paired with the idea of building resilient, regenerative and self-organizing structures, business models and systems to handle complexity and uncertainty, as well as disturbances, shocks and stressors.

As a result, I now believe that the limited impact of Sustainability 1.0 was inevitable. This approach to sustainability lacked both the tools to make fundamental changes at the systemic level, as well as the mindset and narrative required to recognize that sustainable systems go hand in hand with resilience. In many ways, it should therefore not have come as a surprise that we have the two previously mentioned exponential trajectories – one catalyzing exponential growth for capitalism and human evolution, and the other, the exponential degradation of our planet – that now threaten our very existence. Putting it bluntly, Sustainability 1.0 simply didn't have the firepower or tools to successfully change the systemic drivers of capitalism.

Transforming the operating systems that drive Spaceship Earth

Indeed, when we look back over the trajectory that Spaceship Earth has traveled thus far, it paints a pretty depressing picture.

Given this observation, it may come as a surprise to you that my motivation for writing this book stems from my optimism that our species absolutely has the capacity to turn things around.

"Why?!", I hear you ask! Well, my thesis is very simple.

I believe planetary sustainability can finally be achieved if we address the "outdated or malfunctioning code" in our "operating systems". In short, I am referring to the bugs and perverse incentives hardcoded in, what I am going to term as our 'operating systems', that have, to date, prevented the scaling of sustainable products, services, innovations, and the necessary changes in our lifestyle and behaviors.

Removing this problematic code will require the rewiring of everything that sits at the core of what has driven the developed world's economy and way of life for the past 50+ years. [18] Something Sustainability 1.0 has failed to do.

I understand that this sounds like an extraordinary feat to achieve. But nowhere on the cover was it advertised that I was writing a book about "how to easily save the planet"! Sadly, the exponential curves that we are facing are far too numerous and far too advanced in their impact for 'easy' to be part of the equation anymore. But that doesn't mean that transformation isn't possible.

The good news is that digital transformation has already started this rewiring process. Digital technologies offer a measure of hope, precisely because they work hand in hand with systemic transformation and major disruption to business as usual. If you look all around us, digital technologies are already fundamentally changing just about every core 'operating system' that runs our lives or influences our behavior.

According to Peter Diamandis, "*Anything that becomes digitized enters the same exponential growth we see in computing*". Once an activity is digitized in the ones and zeros of computer code, it benefits from Moore's Law (an observation made by Gordon Moore in 1965, referring to an observable process doubling every two years). [19] It therefore has the potential to begin accelerating in an exponential manner.

Our society has become so fast paced and technologically driven that we are now seeing Moore's Law being applied to *everything*, including all living things and our own humanity. Economic transactions, social relationships, education, health, entertainment, news – everything is being upended by digital technologies. Sixty-six percent of GDP was digitalized by the end of 2022. Five billion people are now connected by mobile phones, 4 billion people are being influenced on social media and 2 billion are transacting on digital platforms. [20]

The digital transformation of each sector of the global economy is the single largest opportunity to finally make systemic level changes to the core incentives that drive our behaviors.

This is truly a unique moment in human history, where each of these systems is being fundamentally transformed, reprogrammed and re-coded.

Just in case the above wasn't clear enough, I believe the most important aspect of this process is not the digital transformation of individual products and services, it is the fact that the underlying systems, feedback loops and incentives are also being transformed by it. What excites me even more about the shifts that are taking place, in this regard, is the potential for digital technologies to catalyze a new range of powerful transformational forces that can unleash a global revolution in sustainability for people and planet – if they can be understood, harnessed and directed. To do this, we must ensure that as each of our core 'operating systems' is being disrupted and digitally transformed, their trajectory is hardwired and aligned to our global sustainability goals.

If we look across the 17 Sustainable Development Goals (SDGs) and 169 targets, not one is detached from digital technology. [21] Indeed, 103 of the 169 SDG targets are directly influenced by a combination of seven digital technologies, namely; digital access, 5G, cloud, IoT, artificial intelligence, extended reality (think the 'metaverse') and blockchain. [22]

And the good news is that estimates on how digital technologies can potentially help accelerate action towards our global climate, nature and pollution goals are impressive.

For example, the Global Enabling Sustainability Initiative (GESI) reports that information and communication technologies (ICT) can enable a 20 percent reduction of global CO2 emissions by 2030 when applied to five sectors: mobility, manufacturing, agriculture, energy and buildings. They claim ICT solutions can help cut nearly ten times more CO2 than they emit. [23]

Digital technologies and circular design can also help reduce the embodied materials in goods by 90 percent – through the dual-benefits of efficiency savings and turning products into services in a circular economy. This can significantly reduce the impact of material extraction on biodiversity and the environment. [24] Digital technologies can also help reduce waste and detoxify supply chains by a factor of between 10 and100 times. [25]

I believe that, for the first time in the history of humanity, we have the technological resources and know-how at our disposal to reshape and reformat our planetary operating system and build a circular economy. An economy that supports civilizational sustainability and amplifies the agency of all actors to contribute to sustainability goals, so that we can begin to automate and optimize equally for people, planet and profit.

The choice is now ours to set global level values, goals, timeframes and safeguards for digital transformation. Digital technologies can be designed and deployed in a manner where different values and goals, such as inclusion, equity, regeneration and resilience, can become directly embedded in the underlying code of everything that we build and do. In the process, we would catalyze an enabling environment for our civilization and planet to flourish.

A vision for digital sustainability

The remaining chapters of this book offer a vision and call to action that align with this thesis: a vision capable of uniting the digital transformation and environmental sustainability agendas.

First, as already discussed, I open by arguing that planetary sustainability can only be achieved by transforming the four key 'operating systems' that drive incentives and shape human behaviors. These are the human cognitive systems, social systems, economic systems and governance systems, as well as their relationship to each other. I propose that anyone with a sustainability agenda must understand how digital technologies can influence bugs and outdated code in each of these systems that determine the uptake of a sustainable product, service, lifestyle or behavior. [26] At the same time, we must also connect and integrate our human operating systems with the natural systems that sustain life. Part One of the book is dedicated to unpacking these ideas.

I then move on to the idea that advancing sustainability at a planetary level will only be possible if we learn how to identify, harness and combine what I have coined to be the "Five Forces of Digital Transformation" – Sense-Making, Socializing, Valuing, Embedding and Adapting. I make a case for how these forces will be pivotal for interconnecting the four operating systems, in the process creating incentive structures and feedback loops capable of driving transformation at the speed and scale that our planet needs. I discuss this in Part Two of the book, including examples that showcase the transformational power of each force.

Finally, to truly appreciate the opportunities and risks of applying digital technologies, I then explore how these forces of digital transformation converge to generate exponential transformation. I argue that ideally, digital technologies can help us deploy a planetary operating system that can balance the profit, people and planet equation.

As we no longer have time to wait for linear incremental change, we must now adopt an exponential approach and focus our attention and investments on these highest leverage points for systemic change, but with due consideration that ensures we do not unleash the dark side of these forces. Understanding the full spectrum of opportunities and risks that these forces can generate, and a clear vision for a future where the five forces are harnessed for transformational outcomes, is the focus of Part Three of this book.

Ultimately, however, the hope and vision that has spurred me on to write this book is that you, my dear readers, will feel the urgent call to action to begin more widely using digital technologies to help humanity rewire our 'operating systems'. I hope that you can engage in our shared digital future by considering their application in the design, build and implementation phases of not just your sustainability solutions, but in everything you build going forward, and all that you do. I hope to see our planet reaping the benefits of these solutions, and that they become not only capable of exponential trans-formation, but also of interoperating with other products and services that also form part of the ever-growing digital ecosystem for the planet.

I hope, too, that you will be inspired by how your fellow sustainability practitioners are showcased in this book, and particularly how they are already using the Five Forces of Digital Transformation to direct and shape the evolution of our civilization in more sustainable directions.

And, lastly, I hope that my book will motivate you to begin more actively sharing your innovations, ideas and experiences with your peers, so that best practices can begin to surface and scale more rapidly. I hope for this outcome most of all because I believe that such purposeful collaboration *is* the back-bone of all future outcomes that will ensure every light on our planetary dash-board turns green; for the very first time in our lifetimes.

References

1. Vidal, J. (2012). "Many treaties to save the earth, but where's the will to implement them?" *Guardian*, June 7.
2. United Nations Climate Change. (2021). *NDC Synthesis Report*.
3. NASA. (2020). *2020 Tied for Warmest Year on Record, NASA Analysis Shows*.
4. Waters, C., et al. (2016). "The Anthropocene is functionally and stratigraphically distinct from the Holocene." *Science*, 351: 6269.
5. Steffen, W., et al. (2015). "The trajectory of the Anthropocene: The Great Acceleration." *The Anthropocene Review*, 2: 1, 81–98.
6. Steffen, W. et al. (2004). *Global Change and the Earth System: A Planet Under Pressure*. Springer Science & Business Media.
7. Diaz, S. et al. (2019). *Summary for Policymakers of the Global Assessment Report on Biodiversity and Ecosystem Services of the Intergovernmental Science*. UN Policy Platform on Biodiversity and Ecosystem Services.
8. UN Environment Programme. (2019). *Global Chemicals Outlook II: From Legacies to Innovative Solutions*. Available at www.unep.org, Accessed March 11.
9. World Wildlife Fund. (2014). *Living Planet Report 2014*. Available at www.worldwildlife.org

10. UN Environment Programme. (2014). *Global Chemicals Outlook II: From Legacies to Innovative Solutions.* Available at www.unep.org. Accessed March 11.
11. Arsenault, C. (2014). "Only 60 years of farming left if soil degradation continues." *Scientific American*, December 5. www.scientificamerican.com
12. World Wildlife Fund. (2020). *Will There Be More Plastic Than Fish in the Sea?* Available at www.wwf.org.uk
13. Mohajerani, A., Bakaric, J. & Jeffrey-Bailey, T. (2017). "The urban heat island effect, its causes, and mitigation, with reference to the thermal properties of asphalt concrete." *Journal of Environmental Management*, April 2017.
14. Henley, J. (2020). "Climate crisis could displace 1.2bn people by 2050, report warns." *Guardian*, September 9.
15. Calel, R. et al. (2020). "Temperature variability implies greater economic damages from climate change." *Nature Communications* 11: 5028.
16. Mach, K. J. et al. (2019). "Climate as a risk factor for armed conflict." *Nature* 571: 7765, 193–197.
17. Meadows, D. (2017). *Thinking in Systems.* Chelsea Green Publishing.
18. University of Cambridge Institute for Sustainability Leadership. (2017). *Rewiring the Economy: 10 Tasks, 10 Years.* Available at www.cisl.cam.ac.uk/
19. Britannica. (2022). *Moore's Law.* Available at www.britannica.com
20. Global System for Mobile Communications. (2017). *Number of Mobile Subscribers Worldwide Hits 5 Billion*, Available at www.gsma.com. Accessed June 17.
21. United Nations. (2019). *The Age of Digital Interdependence.* Available at www.un.org
22. Deloitte. (2019). *Digital with Purpose: Delivering a SMARTer2030.* Available at www.gesi.org
23. GeSI and Accenture (2015). *#SMARTer2030: ICT Solutions for 21st Century Challenges.* Available at www.gesi.org
24. Seba, T. & Arbib, J. (2020). *Rethinking Humanity: Five Foundational Sector Disruptions, the Lifecycle of Civilizations, and the Coming Age of Freedom.* Rethink.
25. Ibid.
26. Ibid.

2 Re-coding the 'operating systems' of Spaceship Earth

The four core 'operating systems'

In the previous chapter, we learned about the 'hard' problem of sustainability, namely that our inability to approach sustainability in a way that embraces systems thinking, has prevented us from catalyzing systems level transformation.

In the context of this book, when talking about the need for systems level transformation, I am referring to the transformation of what I consider to be our four core 'operating systems'. These being:

- Human Cognitive Operating System
- Social Operating System
- Economic Operating System
- Governance Operating System

Considered together, these four 'operating systems' are responsible for driving present day capitalism and can broadly explain how the developed human world works. Returning to our analogy of planet earth as a spaceship, we need to ask ourselves if these systems are providing the right kinds of incentives and the real-time feedback that is necessary to manage the essential life support systems of our spacecraft. Natural operating systems are also critically important, but, for now, we will leave them in a mental parking lot – to be addressed in Part Three of this book.

Each of these four operating systems includes specific characteristics, information flows, feedback loops, and time lags that must be navigated in order to successfully design and scale any sustainability solution capable of transformational impact. Looking at these operating systems more closely enables us to identify and consider how vulnerabilities, bugs, outdated code and bad incentives have become 'hardcoded' in each of them.

To solve the hard problem of sustainability, it is critical that these systems undergo significant transformation. And, to achieve that transformation, it is critical to address these design flaws or, as I have termed them, the "bugs and outdated or malfunctioning code". This "code" in each system is largely responsible for creating the perverse incentives that so often undermine

DOI: 10.4324/9781003187523-3

planetary sustainability. More alarmingly, if this outdated code is left to be amplified by digital technologies, it will only further accelerate the destruction of our planet. As observed by Sam Harris – "*It is very easy to believe the worst of individuals until you have some insight into the system in which they are forced to function. If it is a system where the incentives are terrible – even very good, competent and smart people wind up doing seemingly disastrous, stupid, destructive and seemingly evil things*". In short, people will do what their incentives demand. [1]

These bugs also influence how these systems interconnect and influence each other. This is of particular importance, given that these two factors alone are capable of blocking sustainability values and solutions from successfully competing with 'business as usual' products and services. Ultimately, the degree of systems level interoperability designed into a sustainability solution determines its level of scale, speed and transformation possible across each system, as well as at the 'whole-system' or planetary level. [2] To garner a deeper understanding of how these elements impact our operating systems, let's spend some time unpacking them.

Human cognitive operating system

The human cognitive operating system refers to the capabilities of our nervous system and brains to acquire, filter and process the dazzling array of complex information we are bombarded with regarding our internal and external environments. Our cognitive capacities and underlying biology dictate how well we understand this information. This then informs our ability to make decisions and take actions, while also shaping our preferences, beliefs and values. Our human cognitive system forms the backbone of all our other 'operating systems' because human influence and engagement is at the center of these critical processes, driving and shaping them all.

Our cognition works on dual tracks, with each operating at different speeds. One is a conscious 'slow track', involving logic, reasoning and rules, which gear the human brain naturally toward linear thinking. The other, an unconscious 'fast track', involves intuition and the rapid association between a myriad of stored lived experiences, emotions and learned information. [3] All elements of this 'fast track' have been so deeply ingrained within us that their outputs are delivered to our conscious brain for expression on auto pilot. In short, we make snap judgements on hundreds of decisions every day, without ever being fully conscious of them.

These tracks of cognition are shaped by the diverse sources of information to which we are exposed, and by our ability to validate, authenticate and cross-reference that information with our own lived experience, values, cultures and social expectations. Additionally, our individual moods, egos, and conscious and subconscious minds are mapped on top of these systems, together with the impacts of personal relationship dynamics and the natural aging process that add further layers of complexity. [4]

Whilst we have the ability to make conscious decisions, choose our behaviors and exercise our free will, our cognitive operating systems have a number of inherent limitations and bugs which tend to undermine our ability to act, behave and consume in sustainable ways. Let's walk through the major ones.

First, we are limited by the amount of information and the number of variables that we are capable of processing and making sense of simultaneously. This means we are quite poor at predicting interactions between variables and emergent outcomes in systems that are complex, non-linear, ambiguous and uncertain. We struggle with perceiving the cumulative effects of our actions at a planetary scale, or over long time periods, particularly when we cannot directly visualize and/or immediately experience them. These combined bugs are why we struggle in predicting the outcomes of the exponential growth curves discussed in Chapter One. As nuclear physicist Al Bartlett has observed: "*The greatest shortcoming of the human race is our inability to understand the exponential function*". [5]

Consider for a moment the Sustainable Development Goals (SDGs) mentioned in the first chapter of the book. Coordinating the achievement of the 169 SDG targets requires complex analysis, systems level thinking, and optimization for multiple environmental, social and economic outcomes that span local, national and global levels. This is extremely difficult for the human brain to parse given the inherent limitations in our cognitive abilities to process more than a few variables at once. Most people cannot even memorize the 17 SDGs themselves, much less the 169 targets or the variables that impact them. As a result, coordinating collective action at the planetary scale required to optimize and achieve multiple goals simultaneously, and over a multigenerational time span, simply overwhelms our cognitive capabilities.

And if that wasn't problematic enough, these 'processing' limitations are further compounded by other bugs in the human operating system.

In particular, the **second** challenge we face is that our neural biology offers incentives through dopamine rewards that reinforce the immediate gratification of individual needs. This trait tends to prioritize short-term outcomes relevant to our immediate experience, thereby undermining our ability to make decisions, or plan for longer-term action on an individual or collective basis. This is why it is so difficult to form personal habits that can help us achieve sustainability. Most products and services are designed to be simple and convenient as the first order of business, as convenience triggers a dopamine reward faster than complexity. Driving cars, using electricity and purchasing pre-made, packaged food are all convenient aspects of our modern-day life that develop feedback loops for short-term rewards. These convenience incentives often make it far harder for us to make different choices that are more sustainable in the longer term.

The **third** major problem we face is a range of cognitive biases or mental mistakes in our 'fast track' mode of cognition, that can affect our analytical reasoning, logic and rational action. [6] To date, at least 188 of these cognitive biases have been identified that can be grouped into roughly 20 categories. [7] We might think we are acting rationally, but these biases and 'mental shortcuts' are operating in the background, constantly influencing our intuitions and decisions.

For example, many of you, will be familiar with the so-called "attitude-behavior gap" – the idea that consumers say that they are concerned about environmental issues, but do not necessarily express this concern in their market behavior. Confirmation bias is another common cognitive bias that occurs when we selectively collect evidence that overvalues or supports our existing claims or beliefs, while minimizing contradictory evidence. Loss aversion is another bias that involves people being more emotionally sensitive to losing something compared to winning it. A fourth common cognitive bias is our expression of concern, compassion and empathy for the plight of an individual person, while feeling decreasing concern as the numbers expand to a family, community or country. Joseph Stalin once famously observed this phenomenon: "*If only one man dies of hunger, that is a tragedy. If millions die, that's only statistics*". [8] It turns out – we are hardwired to care about one, while filtering out potential concern about many.

And, **finally,** when it comes to processing and acting on environmental risks, we face yet another set of linked challenges. Doom and gloom statistics trigger the emotion of fear and panic, which in turn triggers our flight or fight response. [9] This scenario produces significant physiological and emotional effects on our cognitive functioning.

Physiologically, the blood supply in our brains is redirected from decision-making areas to those that control our limbs. This has evolved so that we can literally run from whatever our perceived risk is, as well as fueling our vital organs so that we can survive. When our brain's blood supply is redirected in these circumstances, our capacity to make critical decisions based on fact, science and sound logic is dramatically impeded.

Emotionally, doom and gloom statistics simply overwhelm most people, undermining their sense of individual agency to contribute to solutions. Such messages are disempowering, and thus counterproductive. The loss of hope – and helplessness – that accompanies climate change can even trigger what has been termed as eco-anxiety and clinical depression, which the American Psychological Association details in a recent report. [10]

If we are ever going to achieve planetary-scale sustainability it is important for us to recognize that these are bugs and not features in the human operating system.

In terms of the ongoing digital transformation of human cognitive systems, digital technologies can now manipulate human behaviors by appealing directly to some of our basic 'fast track' thinking systems, in what has been termed "the race to the bottom of the brain stem". [11] Indeed, our own human agency is now at risk of being hijacked by technology systems that are designed to capture our attention, influence our preferences, and amplify our consumption. This risk already applies to the five billion people who are now online, with this figure set to expand further in the future. In fact, several major tech companies are specifically investing money, effort and resources in growing their reach and market penetration in developing economies and third world nations to ensure this is the case.

As Tristian Harris from the film *The Social Dilemma* likes to say "*Everywhere you turn on the internet there's basically a supercomputer pointing at your brain, playing chess against your mind. And it's going to win a lot more often than not*". [12]

The Five Forces of Digital Transformation, and in particular, the transformational force that I call "Sense-Making", has massive potential. It has the capability to shift the direction on how our human cognitive systems are positively augmented, and even rewired, by these technologies. We must use digital technologies to actively address all four of the major deficiencies mentioned above, if we are to be successful in speeding and scaling sustainability solutions to a planetary level within the next ten years.

Social operating system

Humans are an inherently social species. Our social operating system refers to the network of human-to-human relationships, to the flow or exchange of information and ideas among them, and to how this creates shared belief systems, webs of meaning, social norms and, eventually, cultures.

In every area of our lives, we develop social networks of trusted relationships that we favor to meet our needs. This includes the grouping or networking of individuals around a common structure, such as a family unit, community, church, school, teams, university campuses, company, city, or nation. An individual can be simultaneously connected to multiple social networks at any one time, and the combination often strongly shapes their given identity.

The social system within each group is usually governed by shared values and common identities, myths, and symbols of meaning, as well as hierarchical elements and status protocols that shape the way that human interaction occurs. This includes trust mechanisms, which ensure a particular code, belief system and/or certain behaviors are upheld by all the individuals that make up that group. These mechanisms in turn help determine and coordinate collective behaviors and norms across the group. [13] In fact, in many organizations, informal social relationships between personnel are often as important as formal institutional structures.

Social operating systems are extremely powerful mechanisms for the adoption and diffusion of sustainability solutions and behaviors – if they can be successfully harnessed for such purposes. The impact of social proof and peer pressure on a person's behavior has been widely researched. [14] It has been found to exhibit as much, if not more, influence compared to any other factor. [15] Simply put, ideas, beliefs, preferences and behaviors within a social network are contagious. [16] People are more likely to behave, and even think, in similar ways to their network. This is one of the most important yet misunderstood channels when it comes to achieving global sustainability across the human population of just over 8 billion people. However, regarding the adoption and dissemination of sustainability values, norms and behaviors in our social networks, things tend to go haywire in this operating system in four major ways.

First, we are hardwired to seek out connections with others who are most like us. According to Angela Bahns and Chris Crandall's study on the subject of social behavior, more often than not, future friends or partners are already similar at the outset of their social connection. [17] "*You try to create a social world where you're comfortable, where you succeed, where you have people you can trust and with whom you can cooperate to meet your goals*", Crandall said. "*To create this, similarity is very useful, and people are attracted to it most of the time*". [18] However, surrounding ourselves with likeminded people who provide us with social feedback loops that are predominantly based on that which we already like and believe in, isn't always necessarily a good thing. These kinds of social feedback loops will either reinforce positive social norms and behaviors, or lock in undesirable ideas and behaviors through the development of filter bubbles and echo chambers. [19] When social proof and the collective behavior of our social groups align with our beliefs, identities and/or cognitive biases, it makes it difficult to see other perspectives, recognize blind spots in our own thinking, or change our behavior. Given that the future of our planet and species relies heavily upon our ability to change our collective behaviors, this characteristic poses a real threat to sustainability solutions and outcomes.

Second, many social groups have a strong tendency to reframe or alter facts and narratives whenever the truth or reality of a situation makes them feel collectively threatened or out of control. In these situations, the social group will simultaneously and collectively adopt an alternate belief system and set of behaviors that enables its members to feel safe and in control again, [20] even if, in order to construct this perceived 'safer' version of their reality, they have chosen to make decisions, or base their beliefs, on information that is completely untrue or unverified. As noted by Upton Sinclair, "*It is difficult to get a man to understand something, when his salary depends on not understanding it*". [21]

The dire consequences caused by social groups collectively acting on misinformation or 'alternative facts' have been a pervasive problem throughout human history. Hitler's genocidal campaign against Jews in World War II is but one example. In that instance, the economic hardship of the German state was widely blamed on a single group, even though all evidence pointed to the contrary.

Al Gore also highlighted this phenomenon playing out in our collective narrative about climate action in the documentary *An Inconvenient Truth*. [22] We know that in addressing the climate emergency, we need to make massive changes to the way that our society operates. This means changing lifestyles and economic structures, and so much more besides. This is a scary prospect for a species that inherently avoids change at all costs. Changes to the status quo not only make us uncomfortable, they also threaten entrenched interests and the distribution of power. As a result, we often prefer to turn a blind eye, as the truth is simply too inconvenient to address.

Gustav Le Bon devoted his career to the study of crowd psychology. In his book, *The Crowd: A study of the popular mind*, he states: "*The masses have never thirsted after truth. They turn aside from evidence that is not to their taste, preferring to deify error, if error seduces them. Whoever can supply them with illusions is easily their master; whoever attempts to destroy their illusions is always their victim*". [23]

Third, the dissemination and amplification of information across social networks tends to be weighted in favor of that which will get the most attention and negative emotional reaction, rather than that which is most important, factual or positive. It is easier to trigger and amplify anger and outrage as an emotional response across a social network than almost any other emotion. Stories about violence, hate, fear, division and polarization tend to elicit more of an individual and collective emotional reaction across a social group compared to peace, tolerance, hope, unity and collaboration. [24] Jonah Berger, a professor of marketing at the Wharton School at the University of Pennsylvania, confirmed this phenomenon based on a detailed study in the United States. "*Anger is a high-arousal emotion, which drives people to take action*", he says. "*It makes you feel fired up, which makes you more likely to pass things on*". In short, almost no emotion moves faster across a social network than outrage.

Mass media and political parties have all adopted this negative emotion-based strategy to reach and manipulate the collective actions of their audiences for decades. [25] If it "bleeds it leads" has been one of the dominant ways that media outlets have prioritized the content of newscasts. Fear-based news stories prey on the anxieties we all have and then hold our responses hostage. It is the perfect way to capture our attention and then feed us a barrage of product advertisements, political ideologies and more, that also simultaneously manipulate our sense of security and self-esteem.

This is the same reason why political parties also tend to favor political attack ads against their opponents, compared to positive messages about their own track record and policies. While an ad promoting a candidate's qualifications is more informative, it is the negative ads people remember because of the stronger emotional response that they generate.

This bug also tends to amplify information about our individual vices rather than our virtues – especially when the vices can provoke a strong emotional or even biological response. Brands and influencers have also tuned into this phenomenon since the dawn of marketing. A combination of sex and power works for the same reasons as fear-based news does – it invokes a powerful emotional response that provokes a reaction, capable of amplifying its intensity *en masse*, to market more product and drive revenue. As the saying goes, "sex sells". These forms of crowd manipulation drive populations to want the latest fast fashion fad, iPhone model or Tesla because consuming feels good. The global advertising market is roughly USD600 billion per year[26] – yet expert estimates suggest that only around 10 percent of this budget is aligned to sustainability goals.

On top of these bugs, the media also has a strong tendency to "*collapse systemic issues into personalized narratives*" as observed by podcast host and writer Esra Klein. The media tends to favor and share stories and epic narratives about individuals facing the depths of despair or overcoming herculean obstacles – but struggles with addressing systemic environmental and climate challenges that aren't anchored in individual stories and simple linear solutions.

Due to these bugs in our social operating system, we are now subject to the widespread amplification and manipulation of collective emotions driven by negativity, vices and personalized narratives rather than positivity, virtue and collective action. Information about sustainable products, services, lifestyles and behaviors has a hard time competing for our collective attention because it doesn't invoke a strong enough emotional response to override how we collectively react to that which does. [27]

The **final** bug in our social operating system that tends to undermine the transformation of our operating systems, is the difficulty involved in coordinating collective action between people, organizations and cultures. This involves building trust across social groups and stakeholders, such as businesses, governments, investors and citizens, as well as solving power-sharing dynamics, distribution of resources, sharing of credit and navigating a vast array of cultural norms. Martin Wainstein from the Open Earth Foundation is adamant that "*the greatest barrier to achieving planetary sustainability is not technology – it is our ability to accelerate human trust and collaboration at scale*".

In many cases, collective action is typically undermined in situations where the incentive to 'free ride' on the actions of others is large. [28] This is especially the case in the management of public goods, such as clean air, road infrastructure etc. A free rider, most broadly speaking, is someone who receives a benefit without contributing toward the cost of its production. Once the public good is provided, a person (the free rider) cannot be excluded from enjoying it. For example, if a group of states manage to significantly cut their carbon emissions, other states may not have any incentive to do the same. They will not have to invest in net-zero carbon policies and infrastructure, while still enjoying the benefits produced by those nations who are working toward it. A healthy planet would be the ultimate public good but achieving this clearly involves the organization of a collective action that is nightmarish in proportion.

Another related challenge to collective action is caused by the fact that the perception of risk is a social process. Each culture is biased toward highlighting certain risks and downplaying others. This can make finding consensus across cultures and countries on the importance and magnitude of individual environmental and climate risks difficult to achieve. [29]

As the digital revolution spreads around the world, our social systems have now become a fusion of humans and digital technology. This has laid the foundation for an unprecedented social experiment, in which four billion people are now connected and exchanging ideas through social media, gaming, smartphones and other technology platforms. This offers both powerful opportunities for accelerating social change, as well as vulnerabilities from the bugs that we have discussed in this chapter, not least because communication through these platforms is by orders of magnitude different from anything that we have previously experienced in human history.

The way that digital technologies are used to accelerate the mass adoption of sustainability ideas, behaviors, habits, cooperation and collective action is therefore a critical factor in any plan for planetary sustainability.

However, there is a solution. The Five Forces of Digital Transformation and in particular, the transformational force that I call "Socializing", have massive potential to transform the social operating system. In doing this, they will support sustainability outcomes by addressing all four of the challenges that our social operating systems face.

Economic operating system

In this next section, we will address our economic operating system. This system encompasses how a society chooses to transact, organize and finance the means of production and consumption of goods and services. Ultimately, our economic operating system is concerned with creating institutions, firms, feedback loops and price signals that act as the ultimate matchmaking mechanism between supply and demand, or production and consumption. [30]

Industrial capitalism, as the dominant economic operating system, is based on the idea that markets are the most efficient way to distribute resources across society – based on both competitive and cooperative behaviors and incentives. When we interact with markets, we coordinate with each other a series of collective actions that surpass our individual abilities. Resources and finances are allocated to their most productive use through prices that are determined in markets. [31] Occasionally, market failures such as monopolies, information asymmetries, or externalities require corrective regulation. However, many bugs and perverse incentives have also been embedded into the code of this operating system that work against sustainability.

First, many products and services in the economy depend on the conversion of raw materials and natural resources into finished goods. Our predominant system of production is based on a linear model of extracting natural resources, processing them into value-added goods and services, consuming them and then discarding them as waste. And we have so far failed to place a high value on re-circulating those natural resources back into the economy. [32] Instead, we've built a 'throw away' economy rather than a circular one. We've created products and services with extremely complex 'supply chains' and have failed to accurately measure all the inputs and outputs required.

This tendency means that we often have no reliable way to measure and compare the environmental or carbon performance of a product or service across their lifecycles.

The knock-on effect of this measurement failure is a related failure associated with the production of product or services. That is that we frequently fail to fully account for all the direct and indirect costs that are relevant to the environment and sustainability in general. Our predominant business models and balance sheets often fail to take account of the negative environmental and social impacts that businesses can generate across their supply chains. When possible, businesses also try to externalize these negative costs by passing them on to the public – while privatizing profits. [33] Thus, full cost accounting of the negative and positive impacts of a product, service or firm is still difficult to achieve.

Second, the operating system of the market economy has proven to be extremely effective in terms of matching the supply and demand of goods and services by using a single metric of price to compare value. A single unit of measure was a fundamental breakthrough, allowing the exchange of different products and services against a common benchmark. However, this unit of measure was flawed in that it isn't able to express other qualitative metrics, such as labor standards, ethical principles and environmental performance. In particular, the economic operating system makes almost no attempt to: (i) value the loss of natural capital; (ii) compensate local communities for the inherent loss of value that natural capital represented for them; (iii) prioritize the regeneration of degraded ecosystems or; (iv) place a high value on re-circulating those natural resources back into the economy. [34] Given this, price has now been recognized as a poor metric to calculate and express actual value of products and services. [35] In order to address this, more diversified and holistic methods are needed to measure and calculate value.

This problem also applies to market valuation and stock price of public companies, along with the flow of finance. Today, there is a complete disconnection between the market value of a company and the positive contribution that it makes to society. [36] As economic historian Deirdre McCloskey noted, Western Europe's embrace of the profit motive was the catalyst for what she dubs the "Great Enrichment" of the last two centuries. We know what motivates the market – profit and growth – but now we need to find ways to ensure this same motivation can benefit the planet too. [37] Translating a company's sustainability outcomes so they are reflected in its market value is the missing piece in turning our market economy into our biggest driver of growth for sustainability solutions.

Third, the global financial system is also undermining planetary sustainability rather than contributing to it. This refers to the economic arrangement wherein financial institutions facilitate the transfer of funds and assets between borrowers, lenders, investors and insurance. Its goal is to efficiently distribute economic resources to promote economic growth and generate a return on investment (ROI) for market participants. Ultimately, it is the financial system that controls the flow of capital in the economy and access by businesses to finance their growth. In 2019, the assets of financial institutions worldwide amounted to USD404.1 trillion. [38]

However, because the system focuses on short-term profitability – and lacks metrics on value or environmental risk – it traditionally finances companies whilst failing to consider their impact on natural capital or ecosystem services. The dominant paradigm in making finance decisions has been to focus on financial risk management exclusively, rather than also considering social and environmental risks and impacts.

Financing sustainable development requires capital flows to be redirected toward critical priorities, while divesting away from assets that amplify social injustices, deplete natural capital, and cause climate change. This is one of the single most important levers of change for planetary sustainability. *"If we are to generate truly inclusive wealth then we need a financial system that can efficiently invest in the human, productive and natural capital on which we all depend"*, observed Achim Steiner, former head of the UN Environment Program. [39]

Environment, Social and Governance (ESG) criteria have been developed to help investors screen for potential risks and select companies that meet minimum criteria in these three areas. However, to date, ESG funds only represent USD2 trillion of assets under management – only 0.5 percent of the total financial system. [40] In addition, consistency and standardization in the way that ESG metrics are monitored and calculated is a major problem, and one that has yet to be solved. [41]

Finally, many government subsidies and procurement policies also have the same effect, in terms of creating perverse consumption incentives for damaging sectors that undermine greener alternatives. As was made clear in the UN Environment Programs' 2020 Emissions Gap report, coupling subsidy reform with support for zero-emissions innovations can allow governments to achieve both their economic and their environmental goals. In 2019, the International Monetary Fund (IMF) estimated that global energy subsidies totaled USD5.2 trillion. That amounts to 6.5 percent of the world's gross domestic product in 2017 and was up from USD4.7 trillion just two years previously. If fuel prices had been set entirely by the market in 2015, the IMF indicated that carbon emissions would be 28 percent lower. [42] *"Phasing out environmentally damaging subsidies is a 'no-brainer' and should be the bare minimum that countries sign up to at the talks"*, commented Li Shuo, a policy advisor at Greenpeace China. [43] Similarly, government procurement can have a major influence on market demand dynamics, given that it represents 13–20 percent of global GDP. [44] Public institutions are uniquely positioned to demand transparency to the upstream and downstream impacts of goods and services, and are capable of incorporating sustainability criteria into purchasing decisions at a scale that can shift markets.

The digital innovations of the next two decades will likely alter the essential foundations of our economy and could enable us to coordinate collective action on a global scale. [45] However, if this isn't directed, it could equally contribute to hyper-consumption and exponential levels of resource extraction that will push our planet over the brink. That's why it is imperative that we address the bugs and malfunctioning code of our economic systems.

The Five Forces of Digital Transformation and in particular, the transformational force that I call "Valuing", have massive potential to transform our economic operating system, and to support sustainability outcomes, by addressing these bugs.

Governance operating systems

Governance systems typically involve a series of rules, control mechanisms, and decision-making frameworks that help achieve a system-specific objective. They also typically determine who makes the decisions and sets strategic direction, as well as how stakeholders make their voices heard, how risk mitigation is considered, and how accountability is monitored and enforced. [46] Ultimately, governance is about the distribution of power and influence across groups of people who all share an interest in the outcomes.

Governance operating systems operate at multiple levels and across all sectors. Many people equate governance with governments, as well as the delivery of public services at the level of cities, municipalities, states or nations. However, governance also happens within all types of organizations. It can take the form of governance mechanisms, such as the appointment of a board or oversight committee. But it also has an even more day-to-day application in an organization through the implementation of operational frameworks, internal policies, occupational health and safety and so on. The adoption of financial reporting and third-party auditing mechanisms are other typical manifestations of governance.

Organizations must also consider governance processes and safeguards within the legal context that they operate. In this respect, governance can refer to the management of quality assurance and the mitigation of risk, in terms of product safety standards, data privacy, terms of trade, accuracy of marketing materials, etc.

It should therefore come as no surprise to you that our governance systems have also been designed with many key bugs and vulnerabilities. These have collectively prevented sustainability initiatives to interconnect and scale, especially at systems level.

First, the legislative process and the adoption of laws tends to be slow and rigid, with very little agility or capacity for rapid adaptation. This means our legal frameworks for sustainability lack the ability to adapt dynamically to our changing needs and cannot effectively address the causes of planetary degradation. [47] The very processes of formulating, implementing, monitoring and amending laws in a top down manner is simply too slow to respond to the speed of the economic forces and incentives that are causing environmental degradation. Every step in the chain requires effective feedback loops to guide decision-making, with respect to formulation of public policy, translation of policy into law, implementation of law into action, monitoring compliance and enforcement, identifying positive and negative impacts, and legal amendments. Each step along this impact pathway requires time, resources and capacities that many ministries simply lack. In essence, the companies driving the engine of the economy have more financial resources and agility than the regulatory forces attempting to contain and steer them. Laws always operate at a slower pace than economic growth and the forces of innovation; thus, they are always one or more steps behind effective governance.

Second, our sustainability laws and their enabling environments are often fragmented and siloed solutions meant to address a single problem. This means that they often fail to take account of systems level drivers, solutions, unintended consequences and feedback loops. As one example, a solution such as renewable biofuels in the energy sector creates unexpected impacts in another – forest degradation and food insecurity. [48] Similarly, governments are typically organized into thematic sectors – but rarely do they function in an integrated manner as a whole system, known as a "whole of government" approach. As a result, they often have a limited ability to create any firm or widespread impact.

Third, many governing systems are caught in their own conflict of interest, and therefore operate with a focus on short-term time horizons when it comes to maintaining business as usual. Regardless of whether it is the governance mechanisms within governments or companies, there is often a perverse incentive to maintain the status quo, precisely because disruption causes uncertainty and threatens the continuity of their own future. The leaders of our governments and large enterprises want to stay in their positions of power and authority for as long as they can, and this short-term incentive undermines any long-term structural change that is required to incentivize sustainability.

For governments, this means winning elections and then maintaining the popular support needed for re-election. How many times have we seen political parties campaign on difficult issues to get elected, but then steer around these issues when they are governing, due to the potential pain and social or economic disruption that major reforms may cause? Their political priorities are shaped by those actions that will give their party the best opportunity of being remembered favorably by the public to ensure re-election. This frequently applies to commitments around social issues, sustainable development, and climate action.

When considering how this bug relates to businesses, their modus operandi means ensuring quarterly profitability and growth. A CEO in the business world is legally bound by the company's constitution to ensure the best interests of the company, and that its shareholders are effectively put before virtually any other consideration. Similarly, it is the fiduciary duty of corporate governing boards to focus on maximizing returns to shareholders, rather than considering broader benefits and risks to stakeholders.

Fourth, the prioritization process behind much of our environmental policy isn't about the best policy – it is about how best to allocate and manage resource-constraints and trade-offs in terms of financial resources, infrastructure and human capacities. In short, it is often about identifying what under-resourced areas of a government or business department can afford to do at the time, based on all the constraints they face. Kirk, Reeves and Blackstock argue that *"it is shown that the result [of resource constraints] is that those involved in the implementation process are unable to consider all possible routes to implementation but, rather, only a relatively few 'manageable' options for the regulation of particular activities"*. [49]

This often creates an incentive for path dependency, and a type of "lock in" that favors policy continuation rather than policy change. [50] These terms relate to the significant legacy investment required to put policy into effect in the first instance in both governments and companies. The cost of infrastructure, technology and personnel training are already the 'sunk costs' of a specific policy direction. These costs are usually so significant that there is little desire or political appetite to change focus or direction once a policy is implemented, even if its effectiveness is limited. This generally leads to an inherent short-termism that is inimical to the more sustained policies and action required to address broader and systemic issues. This is why it usually takes some sort of system-level catastrophic failure to occur for any meaningful action to be taken or change in policy to be implemented.

Finally, the dominant governance mechanism for addressing sustainability challenges has been state-centric and top-down; where sovereign governments agree on global goals, and then implement national laws and enabling frameworks to address them. However, what has been missing from this framework is a combination of top-down steering and bottom-up self-organization involving all key actors in the economy.

Through the adoption of laws, there is an implicit assumption that governments have the agency and accountability for sustainability, and that companies and citizens only have the responsibility for compliance. This approach is dangerous for the planet because it undermines the agency of all citizens and companies to directly engage in solving our environmental and climate crisis. By failing to engage 8 billion citizens, our greatest opportunity to shift the needle on climate change will never be realized. Instead, all the power will be invested in traditional governance systems that are plagued by all of the bugs that we have discussed in these first two chapters. It also fails to allow for, or enable, polycentric governance systems – systems that have multiple centers of authority and formality at various scales, rather than a monocentric unit that can be common. In most cases, effective governance is a combination of top-down and bottom-up processes, as well as formal and informal procedures.

While the onus of environmental action has historically fallen on governments, the digital age can change this by enabling the diffusion and decentralization of executive functions to many more stakeholders. In particular, the largest potential impacts of digitalization in governance systems can move in two directions. On the one hand, laws and norms can now be translated into code and algorithms that optimize sustainability outcomes, monitor compliance and dynamically adapt to feedback. And on the other hand, code and algorithms are now effectively driving norms and becoming de facto laws. Increasingly, if a particular variable is not reflected in, or optimized by, an algorithm, it doesn't have 'standing' in the digital space.

The governance implications of this shift are potentially transformative. Sustainability values, goals and commitments can now be hardcoded into the platforms, algorithms and filters of the digital economy at scale – a sort of "SMART" regulation. These values, goals and commitments can then be easily updated, and can also be dynamically adapted to real-time feedback on their effectiveness.

The Five Forces of Digital Transformation, and in particular, the transformational force that I call "Embedding", have massive potential to transform our governance operating system to support sustainability outcomes by addressing all five of the critical bugs that have been outlined in this section.

It's time for transformation

This chapter has hopefully provided you with a much deeper understanding of the bugs and vulnerabilities in each of these operating systems. It is therefore imperative that we understand how to address and squash these bugs, if we are to solve the 'hard' problem of sustainability.

So now, given that you finally have all this background context under your belt, you are ready to meet the Five Forces of Digital Transformation (woo!).

And in doing so, I hope that you can see what I see, and not only gain an appreciation for why I hold these transformational forces in such high esteem, but that you are able to get excited with me as the true scope of opportunities that these forces can create becomes clear. Shall we?

References

1. Sam Harris Podcast. (2021). *#262 – The Future of American Democracy.*
2. Arbib, J. & Seba, T. (2020). *Rethinking Humanity. Five Foundational Sector Disruptions, the Lifecycle of Civilizations and the Coming Age of Freedom..* Tony Seba
3. Kahneman, D. (2011). *Thinking Fast and Slow*. Penguin.
4. Winkielman, P. & Trujillo, J. (2007). Emotional Influence on Decision and Behavior: Stimuli, States, and Subjectivity, in *Do Emotions Help or Hurt Decisionmaking?: A Hedgefoxian Perspective*, eds. Vohs, K. et al., pp. 69–92. Russell Sage Foundation.
5. Bartlett, A. (1976). *The Physics Teacher, Volume 14, Issue 7, The exponential function – Part 1*, pp. 393–394. American Institute of Physics (AIP) Publishing.
6. Winkielman, P. & Trujillo, J. (2007). Emotional Influence on Decision and Behavior: Stimuli, States, and Subjectivity, in *Do Emotions Help or Hurt*

Decisionmaking?: A Hedgefoxian Perspective, eds. Vohs, K. et al., Russell Sage Foundation.

7. Desjardins, J. (2021). *Every Single Cognitive Bias in One Infographic*. Visual Capitalist. Available at www.visualcapitalist.com
8. Doyle. C.C., et al. (2012). *The Dictionary of Modern Proverbs*. Yale University Press.
9. Yu, R. (2016). "Stress potentiates decision biases: A stress induced deliberation-to-intuition (SIDI) model." *Neurobiol Stress*. June 3: 83–95.
10. Clayton, S. et al. (2021). *Mental Health and Our Changing Climate: Impacts, Implications and Guidance*. American Psychological Association and EcoAmerica.
11. Center for Humane Technology. (2019). *Technology is Downgrading Humanity: Let's Reverse That Trend Now*. Available at www.medium.com. Accessed July 18.
12. Thompson, N. (2018). When Tech Knows You Better Than You Know Yourself. Available at www.wired.com. Accessed October 4.
13. Zak, P. (2017). *The Neuroscience of Trust*. Harvard Business Review. Available at www.hbr.org. January–February.
14. Caldini, R. (2006). *Influence: The Psychology of Persuasion, Revised Edition*. Harper Business.
15. University of Southern California. (2011). *Peer Pressure? It's Hardwired Into Our Brains, Study Finds*. Science Daily.
16. Berger, J. (2013). *Contagious: Why Things Catch On*. Simon & Schuster.
17. Bahns, A. & Crandall, C. (2016). "Similarity in relationships as niche construction: Choice, stability, and influence within dyads in a free choice environment." *Journal of Personality and Social Psychology,* 112(2): 329–355.
18. Lynch, B. (2016). *Study Finds Our Desire For 'Like-Minded Others' Is Hard-Wired*. University of Kansas.
19. Arguedas, Dr. A. et al. (2022). *Echo Chambers, Filter Bubbles, and Polarisation: A Literature Review*. Reuters Institute.
20. Harari, Y. (2015). *Sapiens: A Brief History of Humankind*. Vintage.
21. Mukunda, G. (2014). "The price of Wall Street's power." *Harvard Business Review*, 92(6): 70–78.
22. Paramount. (2006). *An Inconvenient Truth*. Paramount Pictures.
23. Le Bon, G. (1895). *The Crowd: A study of the popular mind*. CreateSpace.
24. Shepherd, L. et al. (2013). "'This will bring shame on our nation': The role of anticipated group-based emotions on collective action." *Journal of Experimental Social Psychology*, 49(1): 42–57.
25. Goldenberg, A. & Gross, J. J. (2020). "Digital emotion contagion." *Trends Cogn Sci.* 24(4): 316–328.
26. IMARC Group. (2022). *Global Advertising Market: Industry Trends, Share, Size, Growth, Opportunity and Forecast 2022–2027*. Available at www.imarcgroup.com
27. Lerner, J. et al. (2014). "Emotion and decision making." *Annual Review of Psychology*, June.
28. Hardin, R. (1982). *Collective Action*. Johns Hopkins University Press.
29. Douglas, M. & Wildavsky, A. (1983). *Risk and Culture; An Essay on the Selection of Technological and Environmental Dangers*, pp. 6–15. University of California Press.
30. Mayer-Schonberger, V. & Ramge, T. (2018). *Reinventing Capitalism in the Age of Big Data*. John Murray.
31. Bosch-Badia, M.T. et al. (2018). *Sustainability and Ethics in the Process of Price Determination in Financial Markets: A Conceptual Analysis*. MDPI.

32. Arbib, J. & Seba, T. (2020). *Rethinking Humanity. Five Foundational Sector Disruptions, the Lifecycle of Civilizations and the Coming Age of Freedom.* Tony Seba.
33. Kenton, W. (2021). *Privatizing Profits and Socializing Losses.* Investopedia.
34. Arbib, J. & Seba, T. (2020). *Rethinking Humanity. Five Foundational Sector Disruptions, the Lifecycle of Civilizations and the Coming Age of Freedom.* Tony Seba.
35. Mayer-Schonberger, V. & Ramge, T. (2018). *Reinventing Capitalism in the Age of Big Data.* John Murray.
36. Dyllick, T. & Muff, K. (2016). *Clarifying the Meaning of Sustainable Business: Introducing a typology from business-as-usual to true business sustainability.* Sage Publications.
37. Ranganathan, J. (2021). *Make the Market Work for People and Planet.* January 6. World Resources Institute.
38. Statista. (2022). *Total Assets of Global Financial Institutions from 2002 to 2020.* Available at www.statista.com
39. United Nations Environment Programme. (2015). *New Report Identifies Key Innovations to Bridge Sustainable Development Investment Gap.* Available at www.unep.org
40. Stevens, P. (2021). *There's no Hotter Area on Wall Street than ESG with Sustainability-focused Funds Nearing $2 trillion.* CNBC.
41. Schmerken, I. (2021). *Buy-Side Pushes for Consistent ESG Data and Standard Disclosures.* FlexTrade.
42. Echandi, C. (2021). *To Build Back Better, We Need to Rethink Global Subsidies.* World Economic Forum.
43. Taylor, M. (2021). *Ending Subsidies that Harm Nature Could Create Millions of Green Jobs, WWF Says.* Reuters.
44. World Bank. (2020). *Global Public Procurement Database: Share, Compare, Improve!* Available at www.worldbak.org
45. Mayer-Schonberger, V. & Ramge, T. (2018). *Reinventing Capitalism in the Age of Big Data*, p. 59. John Murray.
46. Zuboff, S. (2019). *The Age of Surveillance Capitalism.* Profile Books.
47. Baumgartner, F. & Jones, B. (1991). "Agenda dynamics and policy subsystems." *The Journal of Politics*, 53: 4 (November), 1044–1074.
48. Clarke, R. (2008). *Biofuels, Climate Change and Food security: Biofuels Position Paper.* December. Tearfund.
49. Kirk, E. et al. (2007). "Path dependency and the implementation of environmental regulation." *Environment and Planning C*, 25: 2, April, 250–268.
50. Cerna, L. (2013). *The Nature of Policy Change and Implementation: A Review of Different Theoretical Approaches.* Organisation for Economic Co-Operation and Development.

3   Introducing the Five Forces of Digital Transformation

Technological versus transformational

I am not the first author to write about the idea that digital technologies are creating a series of forces that will inevitably change humanity. Kevin Kelly's popular book *The Inevitable: Understanding the 12 Technological Forces that Will Shape Our Future* was one of the first to identify specific forces from digital technology, [1] and explain the exponential impact they are having across all aspects of our lives.

His 12 forces include:

1. Becoming: Moving from fixed products to always upgrading services and subscriptions.
2. Cognifying: Making everything much smarter using cheap powerful AI that we get from the cloud.
3. Flowing: Depending on unstoppable streams in real time for everything.
4. Screening: Turning all surfaces into screens.
5. Accessing: Shifting society from one where we own assets to one where instead we will have access to services at all times.
6. Sharing: Collaboration at mass scale. Kelly writes, "On my imaginary Sharing Meter Index we are still at 2 out of 10".
7. Filtering: Harnessing intense personalization in order to anticipate our desires.
8. Remixing: Unbundling existing products into their most primitive parts and then recombining in all possible ways.
9. Interacting: Immersing ourselves inside our computers to maximize their engagement.
10. Tracking: Employing total surveillance for the benefit of citizens and consumers.
11. Questioning: Promoting good questions is far more valuable than good answers.
12. Beginning: Constructing a planetary system connecting all humans and machines into a global matrix.

DOI: 10.4324/9781003187523-4

However, I believe Kelly's analysis didn't go far enough, as he mostly focused on how these forces are reshaping products and services in the economy, rather than how they can be used to propel our common values forward. In other words, Kelly gives no attention to the objectives that can be achieved via these forces for the common good of the planet.

He also stops short in terms of considering how the 12 technological forces could exponentially impact our planet, and what that would ultimately mean for our very existence. In my mind's eye, this is a significant oversight and exactly the gap we need to plug. When examining the trajectory of our technological evolution through the lens of what our planet needs to survive – as well as our own dependency on our planet's survival – it became evident to me that, in fact, there are two overarching archetypes under which all digital technological forces can be categorized.

1. Those that have the power to unleash exponential impact

    scale + speed = impact

2. Those that have this ability, as well as the power, to unleash exponential transformation

    scale + speed + purpose = transformational change

I realized that, to date, almost all technological advancements have been focused on harnessing the power of the first category – as these (and as Kelly points out) address quality of life for the developed world, as well as efficiencies, scale, and, therefore, growth for the private sector. These are also the same set of forces that, at the time of writing, sustainability solutions have typically tried to harness.

As has been established in previous chapters, infinite growth on a finite planet is an unattainable myth and using technological forces to chase after it will only drive us to extinction faster. However, where the hope for shaping a brighter future for our species and planet exists is in how well we learn to harness the forces that fall into the second category. These are the forces responsible for steering the exponential impact generated from other technological forces in the right direction for sustainability. Simply put:

> The Five Forces of Digital Transformation are capable of solving the hard problem of sustainability because they are not only capable of exponential impact across all human operating systems but are also capable of shaping the overall direction or trajectory of these systems, as well as that of all other technological forces at play within those systems.

'How' each transformational force is capable of achieving this shift in direction is contained within the names that I've given them. Let's officially meet each of these forces, for the first time, together.

Force 1. Sense-making

While this force has applications for transformational change across all four operating systems, it can be primarily harnessed to solve limitations in the human cognitive operating system that prevent us from understanding and managing complexity, uncertainty, risk and scale. This includes, but is by no means limited to, measuring environmental, social and economic variables in real-time, understanding interactions and cumulative effects, optimizing their use, and predicting future scenarios. This force helps us measure and make sense of the world. It also helps us peer over the edge of what tomorrow brings, by helping us understand the impacts of exponential growth on our systems and planet, as well as track progress against our local, national, and global goals.

Given that human beings are central to all of our other operating systems, I consider using this force to address the bugs in our human cognitive operating systems to be a critical ignition point. This will enable the force to then have transformational impact, as it diffuses across all other operating systems.

Force 2. Socializing

While this force has applications for transformational change across all four operating systems, it can be primarily harnessed to help sustainability solutions be generated, accepted, diffused and amplified within social operating systems. This is a prerequisite for their eventual adoption in any of the other operating systems. This includes issues relating to trust, engagement, social proof, and the adoption of new social ideas, values and behaviors.

The mass amplification and diffusion of ideas, information, products and services is central to the success of all other operating systems. Therefore, I consider using this force to address the bugs in our social operating systems to be a critical ignition point that enables it to then have transformational impact, as it fires across all other operating systems.

Force 3. Valuing

While this force has applications for transformational change across all four operating systems, it can be primarily harnessed to solve limitations in the economic operating system. We can harness this force to address pricing failures and environmental externalities in the economic operating system. This includes valuing natural capital and ecosystem services, internalizing these costs in accounting and business models, and creating new financial mechanisms and incentives that drive outcomes, such as sustainability, circularity and regeneration. This also includes increasing the social value that we place on natural systems, so that these economic solutions are broadly accepted.

Business models, financial instruments and incentives drive how humans engage with other operating systems, as well as impacting how operating systems interact with each other. Therefore, using this force to address the bugs in our economic operating systems is what I believe to be the catalyst that enables this force to then have transformational impact, as it connects across all other operating systems.

Force 4. Embedding

This force addresses the integration of sustainability values, data, solutions, metrics and outcomes into the codes, platforms, filters and algorithms in our operating systems. By embedding these variables into products, services, infrastructure and systems, we can directly influence, incentivize, optimize and automate the sustainability and governance outcomes that they achieve.

Sustainability data, solutions and metrics are increasingly central to the considerations we make on how to implement technology in all of our operating systems. I believe that harnessing this force to address the bugs in our social and governance operating systems will be the catalyst that enables this force to have transformational impact, as it connects across all other operating systems.

Force 5. Adapting

This force can help accelerate the ability of all operating systems at any level to adapt to changing needs. It achieves this by enhancing transparency, accountability, and decentralization, though it is most certainly not limited to these three processes. This largely involves creating essential feedback loops that can measure the effectiveness of sustainability solutions, policies, behaviors and incentives in real-time. This capability supports the rapid iteration of sustainability solutions, as well as adaptation to constant ecological change, stresses, shocks, disturbances and needs.

The adaptation of all things from analog to digital and digital to exponential, will be central to achieving sustainability outcomes across all our operating systems. Using the force of Adapting to address the bugs in our operating systems is the catalyst that enables this force to then have transformational impact for people, planet and profit.

There's more to these forces than meets the eye

The digital nature of these five forces means that they can also be combined like building blocks, in order to shape the direction of digital transformation outcomes across different sectors, and at the "whole system" level. Once every system is digitally connected, the possibilities to optimize and automate sustainability are nearly endless.

Harnessing these forces for sustainability is an exciting prospect. However, sustainability itself is not the only 'whole system' level outcome that these forces enable us to achieve. Given how rapidly our world is changing, and the convergent crises facing us, we must also intentionally design our planetary operating system to support other emergent outcomes. These include innovation, resilience, agility, regeneration and circularity. These properties go hand in hand with sustainability as key requirements in building regenerative systems, by enabling them to learn, adapt, self-organize and evolve. All these processes are critical for the long-term survival of our species and the health of our planet. But the good news is that our transformational forces have the power to help us achieve this.

I appreciate this vision for digitally enabled planetary sustainability sounds a little too good to be true, but, please, stay with me! I will continue to build this argument step-by-step, force-by-force, throughout Part II of this book, until the bigger picture comes into clearer focus.

Reference

1. Kelly, K. (2016). *The Inevitable: Understanding the 12 Technological Forces That Will Shape Our Future*. Viking.

Part Two

The Five Forces of Digital Transformation

4 Transformational Force #1
Sense-Making

Defining the force

Sense-Making is the process of creating situational awareness and generating insights wherever large volumes of data exist. It also incorporates the procedure of transforming that data into information and actionable knowledge that can deepen understanding, support decision-making and inform action.

Data inputs to sense-making applications can originate from almost anything and anywhere. They include our physical environment, digital platforms, algorithms, datasets, IoT devices and physical infrastructure, as well as people, places and events.

Understanding connections between multiple data points enables deeper insight into the relationships and interactions between the variables inherent in these processes. It also creates more real-time information feedback loops, which can influence and change incentives, decisions and behaviors.

An understanding of the relationships across multiple data sets and variables is essential for the success of any sustainability solution. This applies across the design, build, execution and evaluation phases of any product or service. For example, Sense-Making can **measure**, **monitor and track** the use of natural resources and ecosystems at a planetary scale. It can **filter** and **integrate** vast amounts of data and help **analyze** unprecedented sources of environmental and socio-economic information to extract insights. It can **optimize** the use of natural resources and support spatial analysis of their geographic location and trade. It can also help **model** and **predict** future sustainability trends and risks.

Making sense of an ocean of data

As far as the future of work is concerned, Sense-Making is considered *the* core skill of the 21st century – and it is easy to understand why. [1] The world we now navigate is becoming increasingly complex and ambiguous. Understanding relationships between key human systems has become a paramount need.

DOI: 10.4324/9781003187523-6

Most decisions we are required to make in our personal and professional lives are met by an overwhelming amount of data, not to mention choice. There can often be multiple factors impacting on our decisions, such as benefits and risks across different systems, and various scenarios that require due analysis and consideration. This is a real problem for us humans. As has already been discussed previously, the human cognition system is quite ineffective at carrying out analysis of such data at scale – let alone adding further layers of complexity to the consideration of such information. We simply don't have the capacity to process at this level and scale. And if we think we are overwhelmed by data and choice now; we ain't seen nothing yet.

By the end of 2021, there were approximately 35 billion IoT devices connected to the web. By 2025, this number is estimated to grow to 75 billion. [2] Taking satellites as a single example, there are now over 6,500 satellites orbiting the earth, offering a range of imagery and related analysis. [3] And that number is soon to be well and truly surpassed by the planned SpaceX mega-constellation of 12,000 Starlink satellites, set to be tasked with delivering mass high-speed Internet to the planet. [4]

Today, there is also an average of four devices in existence for every person on the planet, creating more than 50 billion points of connectivity. [5] That number is forecast to grow to an average of 17 devices per person, and a staggering one-trillion connectivity points in the next five years. [6]

So now that humanity has this new-found capacity to engage at this level of connectivity, our evolution from the Internet of Things (IoT) to what is now being referred to as the Internet of Everything (IoE), is all but certain. In our lifetime, we will bear witness to a level of connectedness that will see this global network exponentially expand before our very eyes. In short, there is a digital swarm of sensing devices fast approaching and in plague proportions.

This swarm will digitally index, monitor and measure our natural resources, every aspect of our environment, and much of our physical infrastructure. This will extend beyond the external world to be used to map, monitor and measure the most microscopic inner workings of our physical selves, our personal movements, interactions, transactions and even our behaviors. And for what purpose? Quite simply, data.

The natural by-product of having so many instruments of measurement and points of connectivity at our disposal is the staggering amount of data, aka Big Data, that is now being generated. So how *big* is big?

By 2025, humanity is predicted to produce around 175 zettabytes of data per year (1 zettabyte is 1 billion terabytes). If this data was converted into DVDs and lined up end-to-end, it would circle the earth 222 times, or be equivalent to 23 trips to the Moon. [7]

With the prevalence of so much data, it is no wonder humans need to harness Sense-Making as a skill. This human necessity, combined with our limited capacity to understand, protect and process such a magnitude of data, has led to the emergence of the transformational force that is Sense-Making.

This force is made up of technologies such as artificial intelligence, machine learning, cloud computing, blockchain, digital twins, and many related innovations.

These technologies are already playing an essential role in our sustainability toolkit because they can be applied in immeasurable ways to each of the three levels of digital transformation. Let's discuss them here:

Level 1 (Products/Services) focuses on how to make a sustainable product, service or value-chain more efficient and effective for the end user. With respect to the force of Sense-Making, this refers to real time measuring, monitoring and managing of individual scenarios, organizations and solutions.

Level 2 (Brakes and Accelerators) focuses on how stakeholders such as governments and organizations can work to address brakes and accelerators in the systemic adoption of a sustainability product, service or value-chain. With respect to the force of Sense-Making, this refers to transforming insights and intelligence to inform feedback loops to accelerate sustainability at systems level.

Level 3 (Digital Ecosystems) focuses on how organizations and their products, services and value-chains can contribute to digital sustainability ecosystems and whole of society transformations. With regards to the force of Sense-Making, this refers to structuring information flow, feedback and foresight to drive whole system outcomes

The following case studies will explore how the force of Sense-Making has been revolutionizing sustainability across these levels and a range of sectors. Together, we shall walk through their journeys as they deployed Sense-Making to solve specific problems and scale their sustainability solutions.

Integration of data for insights at local level

Farmers and ranchers have been dogged by drought, floods and bushfires for decades. However, this past decade has seen climate change escalate these hardships to unprecedented levels. Hitting this crisis point was the final straw that has pushed an historically conservative industry into embracing a combination of Sense-Making technologies, such as cloud computing, remote sensors, drones, satellites and AI.

This application of Sense-Making technology in the agriculture industry is now so widespread that it has been given a specific term and is now referred to as 'precision agriculture'.

Precision agriculture is an approach to farm and ranch management that uses digital technologies to ensure that crops and livestock are given an optimum amount of nutrients and access to water that will help them to thrive maximally. It can also help farmers to understand weather patterns and predictions, in order to inform crop planting and harvesting decisions. Precision agriculture techniques can accurately measure and monitor valuable commodities, such as fuel, fertilizer, pesticides and water, while automating the application of inputs for such things, leading to more efficiency and less waste. It can also help farmers and ranchers determine and predict commodity prices and identify market opportunities.

Precision agriculture helps farm/ranch owners and managers achieve these things by providing access to real-time data and dashboards, which are relevant to their chosen crops, livestock and farming/ranching methods. This means that they have better control over the outcomes on their farm or ranch, can mitigate risks in a more time- and resource-efficient manner, and can make more accurate decisions. Ultimately, these capabilities mean farmers and ranchers can run far more efficient and profitable businesses that also make significantly less impact on the environment.

One of the dynamic scale-ups that is trying to use digital Sense-Making tools to solve a key challenge for the farming of commercial livestock is Farmbot (also known as Ranchbot in the USA). The goal of this company is to provide farmers (or ranchers, as farmers of livestock are referred to in the US) with a smart, cost-effective and user-friendly solution to better manage their most valuable and scarce resource – water.

Farmbot was established in Sydney, Australia in 2014 and is the brainchild of Co-Founder and Chief Operating Officer, Craig Hendricks. Craig is all too familiar with the obstacles and hardships facing Australian livestock farmers. He started developing Farmbot after dealing with a range of difficult challenges in managing his own property. These left him looking for smarter ways to monitor and manage the water resources on his land. And it didn't take long for his ideas to spread. Millions of acres of land around the world are now equipped with Farmbot's water monitoring solutions to keep track of their water resources.

Anna and Andrew Cochrane are also no strangers to the high labor, fossil fuel and cost-intensive practice of water monitoring. They manage Isis Downs Station for Consolidated Pastoral Company (CPC), a 237,000-hectare property situated 130 kilometers southeast of Longreach in outback Queensland, Australia.

They can run anywhere from 18,000 to 26,000 head of cattle, depending on the season. They have 16 staff members on the property, one of whom is employed full-time to drive approximately 1,200 kilometers each week (62,400 kilometers or 74,880 miles, annually) to manually monitor and maintain 180 livestock watering points on the property. There were also historically two aerial inspections carried out on the water points each week.

In 2018, Isis Downs Station began trialing Farmbot remote water monitoring sensors, and now has 20 such sensors on their property at strategic waterpoints. Implementing Farmbot technology has reduced the financial impact, number of staff hours, and amount of fossil fuel required to perform this function by about 33 percent. A real game-changer and adding additional sensors would drive down the cost further. These technologies have also paved the way for access to real-time monitoring of their most important water sources, so they can respond more quickly to emergencies with flow levels, accidents and leakage. The value added by Farmbot has since led to a systematic program of installing sensing units on many CPC properties.

So how is Farmbot utilizing Sense-Making to achieve this?

Farmbot uses a combination of internet and/or satellite-connected remote sensing devices, satellite imagery, cloud computing, and platform technologies to offer near real-time monitoring and custom analytics to farmers and ranchers. On a typical farm, the company installs as many wired sensors as a farm has water sources.

This data is then fed into their Software as a Service (SaaS) platform, putting farmers 'in the know' on their water assets, through using a dashboard which they can access from their desktop, tablet or mobile phone 24/7. Farmers are notified through this dashboard of any significant and/or rapid changes to water levels, water quality, or any of its delivery system/lines via a series of alerts. These real-time insights also generate reports on water trends, consumption, quality, system/line faults, leak detection and abnormalities.

Farmbot has also incorporated weather station capabilities into their sensor technology. This enables users to gain access to their disaggregated daily rainfall for specific locations on their property, enabling them to understand local micro-climates and associated crop suitability. Farmbot has also worked to address some of the key brakes and accelerators in scaling sustainability, by putting a more accurate price on the cost of water inputs and loss.

The software also has more macro applications as well, with Farmbot also exploring how their offering can enable other devices to connect-in with their platform to form part of a wider digital ecosystem. This will help support broader sustainability outcomes. For example, by linking up on-farm water analytics – with predictive models on climate change risks – primary producers can then more sustainably manage precious resources, and better understand longer-term risks, trends and options.

In the not too distant future, I could well see these kinds of technologies also capable of connecting anonymized data into other platforms capable of providing insight to consumers, industry partners and regulatory bodies to provide greater transparency around the supply chain, support more sustainable use of resources and arable lands as well as maximize food production.

Processing of big data for global insights and planetary-scale dashboards

Of course, Sense-Making of water distribution and availability at the farm level, isn't the only killer app. The power of Sense-Making with cloud computing and AI is that it easily scales to the global level as well.

One of the Sustainable Development Goals requires countries to report on the extent of their surface fresh water (SDG 6.6.1). This is a critical indicator to monitor at the global level in the context of climate change, increasing freshwater scarcity and global food security. However, many countries simply don't have the capacity to achieve nationwide monitoring due to a lack of key infrastructure, including water monitoring gauges and weather stations. Indeed, water monitoring at a national scale can be costly for massive countries, such as the US, Russia, China, India, and other highly populated nations.

Since 1984, the global archive of free satellite data from NASA has accumulated 3,066,102 Landsat images – over 1.8 terapixels of data. Do you want to take a guess at how long it would take a single computer to process this data to extract surface water insights?! The answer is approximately 6 million hours of computation (over 1,212 years) to complete. As the head of Google Earth Engine, Rebecca Moore likes to say, "*This is equivalent to powering up your computer just after Charlemagne conquers Saxony in the year 804 and then leaving it running 24 hours a day, 7 days a week … and then water maps just might be ready today*".

This is where the cloud computing coupled with the AI of today has made exponential leaps and bounds. Today, this same task can be accomplished in just 45 days using a cloud of 10,000 networked CPUs and machine learning algorithms. We now have unprecedented processing power to monitor the vital signs of the Earth at a planetary scale.

To make this kind of critical water data available to the world, the European Commission's Joint Research Center (JRC), the Google Earth Engine, and the UN Environment Program teamed up to develop the sdg661.app, and named it Global Water Explorer. It shows changes from 1984 to the present day through interactive maps, graphs and full-data downloads, providing critical statistics for every country's annual surface water (such as lakes and rivers).

The new app aims to make this water data open, freely available, and easily accessible to everyone. Countries and water sector stakeholders can compare data with one another, and, for the first time, gain greater understanding of the effects of water policy and infrastructure, such as dams, diversions and irrigation practices on water bodies that are shared across borders.

For countries that have never had this information, the app provides free, scientifically validated data, updated annually, that can now inform their environmental policies. This provides a compelling, unprecedented, and globally consistent way of measuring surface water and changes over time.

Of course, one of the main challenges is now combining this analysis with other sources of data, so that it can inform, for example, farm-level decision-support. At the moment each pixel on a landsat image represents 30 m², and the satellites only pass above the same place on Earth every 16 days. This represents a barrier to uptake – so the next key step is offering additional on-demand analytics through other satellite providers, such as Planet.

Planet offers daily imagery of every inch of the Earth's surface at 3–5 meters resolution; meaning you can see images of every part of the Earth's surface as if you were standing 3–5 meters away from it. For surface water monitoring this is an absolute game changer! Planet's technology has the capability to provide detailed imagery of existing global surface water sources which can then be compared to historical data to identify key trends that can be used for filtering potential problems. In the not too distant future, it is easy to see how this level of data can then be overlaid with other platforms like Farmbot for instance to provide even higher levels of insight and prediction.

And yet, in spite of the advancements we have made with the transformational force of Sense-Making, we still have to overcome another key challenge in developing a digital ecosystem of monitoring platforms. At the moment, we have an interoperability challenge, in that the SDG 6.6.1 platform, the Planet platform, and platforms like Farmbot have no standard way of communicating, nor can they share data between themselves and any other random platform or technology without significant investment and intentional design.

However, water monitoring isn't the only example of where interoperability between solutions, platforms, and data, on a global scale, is needed. There are now dozens of amazing examples around the world of digital platforms that conduct regular monitoring of environmental change. From illegal fishing with Global Fishing Watch, to global scale deforestation with Global Forest Watch, these platforms can now make sense of huge datasets, and provide analytics that can inform real-time decision support, stopping illegal activity in its tracks. Microsoft has also recently launched its Planetary Computer initiative, which puts global-scale environmental monitoring capabilities in the hands of scientists, developers, companies, and policy makers, enabling data-driven decision-making for global operations.

With so many incredible Sense-Making applications already in existence, imagine the level of insights and progress toward a sustainable future that could be made if all these platforms, and the data that they collect, were able to communicate with one another.

Integration of big data and predictive analytics

One of the major challenges in the development of renewable energy is the unpredictability of weather patterns. This makes it difficult to predict sunshine and wind levels and tap into alternative sources during peak power demand.

IBM has developed a machine learning algorithm that can absorb information about weather from thousands of data points, and predict days, even weeks, in advance how much power from solar and wind farms will be available for the US power grid.

The new system is as much as 30 percent more accurate than today's state-of-the-art weather prediction systems – used by organizations such as the National Weather Service – according to the National Renewable Energy Laboratory.

The system works by combining more than 1TB of data gleaned daily through more than 1,600 weather-monitoring stations and solar and wind plants in the continental US, as well as from weather satellites. [8] This data is overlaid with data sets comprised of images from cameras mounted at wind and solar farms that watch the sky for weather patterns. The great thing about this kind of AI-powered platform is that the more data you put in, the smarter it gets.

Because the system can better predict how much renewable energy will be available, the nation's power grids are better able to integrate that electricity with traditional forms of power. As the amount of solar and wind capacity continues to grow, it's critical for regional power grids to know how much renewable electricity they'll have in advance to better plan their capacity needs.

This forecasting application by IBM is an example of a digital twin, where a complex process is represented and modeled within a digital environment. Digital twins of ecological systems and natural resources are a massive application of Sense-Making. In accordance with this, the European Union is building an entire digital twin of the planet, named Destination Earth. This will enable planetary scale modeling of major environmental challenges, such as extreme events, climate change adaptation, protection of biodiversity, and modeling of pollution.

Verifying, modeling and decisioning

As previously established, more data has been produced in the last two years than in the entire previous history of the human race and thanks to our growing population, multiplied by the exponential growth in digital technologies, this is only the tip of the iceberg. So, it isn't just the amount of data that poses challenges for sustainability but being able to authenticate its source and quality also. Technologies that can verify data hold the key to our ability to trust sustainability claims a company makes or the sustainability intentions and impact of different people and organizations.

The climate-trace coalition is a first-of-its-kind environmental analytics tool, backed by an international coalition that includes seven environmental nonprofits, and former Vice President Al Gore. It uses a combination of machine learning, infrared satellite imagery, and advanced computer modeling to track polluters worldwide in real-time. The information it collects, accessible through a data-rich online portal, will help environmental organizations verify that governments around the globe are honoring pledges to cut greenhouse gases. The data will also enable companies to better judge their supply chain's cleanliness, and help the public stay informed.

These three applications of Sense-Making are enabling the public to obtain more detailed information about the integrity of their governments' pledges and the origin and chain of custody of different products. But Sense-Making can also help companies calculate the full environmental or carbon footprint of a product across its full supply chain.

Embedding digital technologies such as RFID tags, IoT sensors, or blockchain applications across a company's value chain, or supply chains, offers transformational opportunities to systematically collect detailed data on the sustainability journey and chain of custody of a product, from raw material to market.

This data can then be processed using lifecycle assessment (LCA) methods, or Product Environmental Foot printing (PEF) to accurately measure the environmental performance of different products, and the role of different players in the supply chain. LCA and PEF can help identify major priorities across the supply chain for impacts, enabling us to better design with the end – or rather 'next life' of a product – in mind from the perspective of the circular design of goods.

One of the most difficult challenges has been accurately calculating the total greenhouse gas emissions (GHG) of a product across the three different scopes of their operations. These are typically divided into:

Scope 1: covers direct emissions from owned or controlled sources.
Scope 2: covers indirect emissions from the generation of purchased electricity, steam, heating and cooling consumed by the reporting company.
Scope 3: includes all other indirect emissions that occur in a company's value chain.

Initiatives such as Carbon Mark are trying to use Big Data and Machine Learning to automate the calculation of carbon emissions of all consumer goods, incorporating their full-lifecycle, and enabling the ranking of comparable goods accordingly. The Carbon Mark is primed to create the capability to instantly calculate the carbon footprints of all consumer products worldwide. Moreover, the ambition is to make Carbon Mark an open-source international standard and digital public good, so that it can be applied by any company to measure their supply chain performance.

A first step to enable product comparability and digital nudging, a concept we explore more fully in Chapter Five.

Salesforce has also entered this space, launching their new SaaS offering, Net Zero Cloud 2.0 in February 2022 which takes the pain out of data management. It enables real-time, accurate carbon emissions data dashboards, provides stakeholder transparency and supplier relationship management. Finally, it leverages great benchmark data and facilitates scenario planning, via the ability to create simulations to see how different decisions or investments will impact your carbon footprint, informing the process an organization takes to set emissions targets accordingly.

JetBlue became the first airline to use Salesforce Net Zero Cloud for emissions tracking to aid them in accelerating their journey toward sustainable air travel. Following the implementation of Net Zero Cloud, JetBlue will make travel emissions data available to the airline's Sustainable Travel Partners ensuring that carbon emissions data relevant to their ecosystem is not only available, but usable across their full value chain.

"We're excited for Net Zero Cloud to not only help us strategize and plan for our own 2040 Net Zero target, but to also help us share that data with our Sustainable Travel Partners, providing first-of-its-kind emissions reporting based on each corporate traveler's actual JetBlue flights. As the world comes together to reduce our collective environmental impact, collaborations like these with Salesforce are vital as we share expertise in pursuit of shared goals."

(Sara Bogden, Director of Sustainability and
Environmental Social Governance, JetBlue)

And as we move into a world where governments are requiring actors to provide information with an increasing level of transparency regarding their Environment, Social and Governance (ESG) data, every organization will need to have a strong story backed up by evidence of their sustainability track record as well as proof of their commitment to act responsibly for our people and planet.

IV.AI is one such company helping ESG leaders avoid red-flags, track scientific advancements, and compare peer strategies at scale. Global leaders benefit from a live feed of all ESG reports, web posts, papers, podcasts, books, and social conversations from experts, with AI that has been trained on all the ESG data so anyone can compare approaches, find anomalies, and reduce blind spots. Their algorithms map the whole market so ESG leaders can better understand the ESG frameworks, commitments and progress of their competitors, public companies and ratings data, and identify opportunities they aren't currently accessing.

The solution helps ESG leaders reduce risk by better tracking what every other ESG leader is delivering within their sector and industry and by comparing those classifications versus red flags being introduced from academia, and influential public pundits. Their models can also support decision makers in identifying blind spots in their ESG strategy that leave them open to stakeholder and customer scrutiny as well as potential breaches of government regulation.

The IV ESG solution showcases all the specific constructs and topics present in all ESG reports so you can reverse engineer ratings providers with an easy-to-use toolkit that maps to taxonomies used by Sustainalytics, European Financial Reporting Advisory Group (EFRAG), Sustainability Accounting Standards Board (SASB) and more.

Organizations accessing these types of technologies can choose to further amplify their impact by combining the data and insights they derive with other applications designed to inform national-level policy making around supply chains, namely the Sustainable Consumption and Production Hotspot Analysis Tool (SCP-HAT). This platform calculates the consumption footprint of a country by adding the pressures and impacts related to imports to those occurring domestically and subtracting those related to the exports. It analyzes key indicators in a country's economy to identify the 'hot spot' areas of unsustainable production and consumption, in order to support priority setting. The SCP-HAT allows for analyzing direct as well as indirect impacts brought about by the production and consumption activities of national economies and international trade. It is, therefore, able to identify hotspots related to domestic pressures and impacts, as well as impacts occurring along the supply chains of goods and services for final consumption in any given country.

Recently, the UK government has moved to implement trade embargoes on carbon-intensive imports as part of their ambitious Net Zero strategy. Technologies such as SCP-HAT will become increasingly important in the implementation and accounting of such measures at scale.

In fact, many governments stand to benefit from the use of Sense-Making for supply chain analysis and policy making at the national level. For example, the ARIES platform (Artificial Intelligence for Environment & Sustainability) offers an AI-driven modeling platform for natural capital accounting. It is designed to support a System of Environmental-Economic Accounting (SEEA) by identifying all of the agents involved in the nature/society interaction, connecting them into a flow network, and creating the best possible models for each agent and connection. The result is a detailed, adaptive, and dynamic assessment of how nature provides benefits to people and the economy – which has the potential to be nothing short of transformational.

Summary

Through the above examples, I hope to have given you a glimpse of the potential for Sense-Making to overcome the limitations of the human cognitive system in order to revolutionize environmental sustainability. If the force of Sense-Making starts to scale for environmental sustainability, we will see massive transformational outcomes, including the following:

First, real-time collection, verification and validation of the provenance, quality, accuracy and consistency of data and its sources on the state of our planet, and the ability to measure the effectiveness of different actions taken by the private and public sector as well as civil society.

Second, real-time environmental and carbon performance information will be available for every company, product or service on the market. We will achieve full transparency of the environment and carbon performance across a product's supply chain. It will also be impossible to ignore or hide the ways that companies help or hinder the implementation of the SDGs.

Finally, Sense-Making will enable all the individual data sources to interact, interoperate and intersect with each other either as a whole or in some way collectively. We can begin to integrate and model different systems within digital twins, in order to understand trade-offs and develop simulations for improved risk mitigation and response. This will be the catalyst to shift us away from linear thinking and decision making. We will be able to reach into the future to see beyond the immediate and short-term ramifications of our decision making to understand long-term and second, third and fourth order consequences of our actions and how they impact on the generations to come.

However, for Sense-Making to continue to grow as a transformational force, some enabling conditions are also needed.

The first involves deciding on the criteria that should be used to evaluate sustainability; acquiring the required performance data; and enabling comparability in sustainability performance across different organizations.

The second involves managing the input data as digital public goods, and ensuring a governance framework that protects data provenance, quality, accuracy, interoperability, and privacy, as well as establishing safeguards and international standards for public-private partnerships. In the short term, the public and private sectors need to agree on the core high-value data sets and standards that are required to underpin strategies, for both countries and companies to achieve the SDGs and multilateral environmental agreements (MEAs). These will need to be quality controlled, interoperable, and financed with a sustainable business model.

This is why developing a set of Earth API standards and interoperability frameworks is so critical. APIs, or Application Programming Interfaces, enable digital platforms to talk to each other, and share data in a standardized format. These would enable different platforms to seamlessly share data and analytics, as part of a digital ecosystem of data for the planet. In addition to an API framework for environmental data, a digital ecosystem for the planet also needs a simple way to measure and communicate the quality of each data set, the level of uncertainty, the suitability to be used in different kinds of analyses, and a standard digital object identifier (DOI). This latter DOI element is essential for it to be possible to track how different data sets are being used and commercialized, and how cost-recovery business models can be created to pay for these essential digital public goods.

Another enabling condition involves building trust in the data and technologies with stakeholders and ensuring transparency in the underlying algorithms. Algorithmic transparency is an essential precondition to digital trust, enabling potential sources of bias to be detected. Both regulators and the public need to be able to see how different algorithms are being optimized for sustainability outcomes, while balancing profit, people, and planet.

Sense-Making is an integral part of many different aspects of civilization, such as modern science, engineering, everyday life and more. The ongoing digital revolution has provided us with an extended ability to measure phenomena at many scales, in a way never before dreamed possible.

References

1. Co:Lab4. (2019). *Sensemaking, the Core Skill of the 21st Century.* Available at www.medium.com. Accessed March 7.

2. Security Today. (2020). *The IoT Rundown For 2020: Stats, Risks, and Solutions.* Available at https://securitytoday.com. Accessed January 13.
3. Mohanta, N. (2021). *How Many Satellites Are Orbiting the Earth in 2021?* Geospatial World. Available at https://geospatialworld.net. Accessed May 28.
4. Howell, E. et al. (2022). *SpaceX Starlink Internet: Costs, collision risks and how it works.* Available at https://space.com
5. Lant, K. (2017). *By 2020, There Will Be 4 Devices for Every Human on Earth.* Futurism. Available at https://futurism.com. Accessed June 19.
6. Diamandis, P. (2015). *The World in 2025: 8 Predictions for the Next 10 Years.* Singularity Hub. Available at https://singularityhub.com. Accessed May 11.
7. Coughlin, T. (2018). *175 Zettabytes By 2025.* Forbes. Available at https://forbes.com
8. Mearian, L. (2015). *IBM's Machine-learning Crystal Ball Can Foresee Renewable Energy Availability.* Available at https://computer world.com. Accessed July 17.

5 Transformational Force #2
Socializing

Defining the force

Socializing is the process of increasing the awareness, reach and adoption of an idea, behavior, product or service. Socializing impacts the rate of adoption, as well as the reach of acceptance for sustainability solutions, and is therefore a fundamental force for systems transformation. Socializing uses a combination of social proof and network effects to spread information across applications and systems that connect and/or influence large numbers of humans.

The transformational force of Socializing has many exponential properties. It can be used to **engage, communicate and immerse people in data and stories** that influence their emotions and behaviors. It can help build **acceptance** and **trust** in a product, service or behavior by showing how 'influencers' and peers in a social network are adopting it. It can **nudge consumers** towards products, services, behaviors and actions that have the best environmental performance by offering social proof of popularity. It can offer social and economic **incentives and positive reinforcement** for specific behaviors by offering recognition, rewards and immediate feedback. It can help consumers **share** information, products and services as part of the **collaborative** economy. Socializing also underpins the process of collective **learning, social resilience,** and **collective actions** such as citizen science and crowdsourcing.

From a digital sustainability perspective, socializing is such an essential force in our toolkit because it can be applied at the three levels of digital transformation.

Level 1 (Products/Services) Enhanced awareness, trust-building and incentives for exploring new information, products and services.
Level 2 (Brakes and Accelerators) Creating wide-scale adoption of new behaviors, enabling collaborations and sense of community and global diffusion of information, ideas, products and services.
Level 3 (Digital Ecosystems) Creating feedback loops that reinforce and reward sustainability behaviors, habits and lifestyles and enable new and sustainability focused business models.

DOI: 10.4324/9781003187523-7

The rise and influence of social media platforms

Today, we are running a massive global experiment in terms of the level of social connectivity that is now possible through digital channels and social media. At the same time, nearly every transaction in the economy is being mediated through digital algorithms and code. The combination of these two developments means that there is a massive opportunity to use these entry points to socialize sustainability solutions through these digital channels.

Today, more than 5.2 billion people own a mobile phone, 4.5 billion people use the internet, and social media users have passed 3.8 billion. [1] At the time of print, 65 percent of global GDP has been digitalized. Fast forward a further ten years to 2032, and it is estimated that 70 percent of new value created in the economy will be based on digitally enabled platforms. [2]

Big numbers, yes, but what do they mean? Let's think for a moment about the impact that this sea of devices has on our lives. Take our social relationships for example. Nearly 60 percent of the world's population is already online. Just under half the population don't just own mobile phones, they instead own smartphones, and use them to access social media, spending an average of 3.3 hours per day on their screens. [3]

Mobile apps now account for 10 out of every 11 minutes we spend using mobile devices, with web browsing only responsible for 9 percent of our mobile time. [4] Roughly half of the time that people invest in using mobile phones each day is spent using social media and communications apps. [5] The world's smartphone users downloaded more than 200 billion mobile apps in 2019, with these generating a total of USD120 billion of revenue through initial and app-related purchases. [6]

On August 27, 2015, Facebook reached an incredible milestone: 1 billion simultaneous users on the network. [7] Today over 2.5 billion people are on Facebook, [8] and many other platforms are not far behind. Youtube has 2 billion, [9] Instagram has 1 billion, TikTok has 800 million, Weibo has nearly 500 million, Reddit has 430 million, Twitter has 330 million and Pinterest has 322 million. [10] Gmail dominates online email services with 1.5 billion global active users. [11]

Not even iconic sci-fi movies like *Back to the Future* could have predicted the voracity with which we now engage in this social experiment – this facet of our lives has now become completely unrecognizable to the uninitiated.

Using technology for our social engagement is but a single, tiny example of how radically different everything that has been touched by the transformational force of Socializing has become. As technology continues to infiltrate all aspects of our lives, and the world around us, we need to accept that any plan we wish to have for our future will be largely determined by this transformational force.

Hacking our own sustainability behaviors at scale

As discussed in Chapter Two, we humans have many 'bugs' in our own operating systems that tend to disrupt our ability to take specific decisions and actions that would advance planetary sustainability.

We prioritize our immediate needs, and discount the future, in terms of needs and consequences. We have a hard time conceptualizing geographic scales at the planetary level, or timescales involving unidentified future generations and global civilizations. We tend to think in linear pathways, rather than complex or exponential concepts. In other words, we have short memories and get easily distracted by bright shiny objects. We are infinitely adaptable to new levels of normal and shifting baselines. We can hold contradictory views in our mind (cognitive dissonance), and we tend to favor information that is aligned with our existing beliefs (confirmation bias). Our actions stem from both logic and emotion, mind and spirit, reflection and automation. Indeed, we are complicated beings.

These bugs will undermine our collective sustainability if we can't figure out how to use digital technologies to rewire these circuits or bypass them completely and at 'whole society' level. We simply need to hack our own sustainability behaviors (of course, in an ethical manner). Indeed, although many consumers would like to buy more sustainable products, only 20 percent do so regularly. [12] This is also known as an "attitude-behavior gap", since the attitude towards a certain behavior is often a poor predictor of the actual marketplace behavior. [13] Fortunately, new techniques are being explored that we can add to our toolkit immediately.

New insights from behavioral science suggests that consumers can be guided to make sustainable choices, without communicating explicitly about sustainability goals. In particular, the concept of nudging is about designing the landscape of choice that a consumer is faced with, using non-intrusive interventions that guide people toward a desirable action without limiting their choice. [14]

In the context of sustainability, The basic idea of nudging is to make the unsustainable decision harder, and the sustainable decision easier to make. Although striving toward the same goal – establishing a desirable behavior – nudging is distinguished from laws and rules. Instead of being forced, nudging enables an individual to have freedom of choice.

Nudge theory was conceived by Nobel Prize-winning economist Richard Thaler and Harvard Law School professor Cass Sunstein in 2008. Drawing on behavioral economics and choice architecture, nudge theory proposes that positive reinforcement and indirect suggestions can influence the behavior of groups and individuals. For example, to persuade shoppers to eat better, a supermarket would display its healthiest food at eye-level on store shelves. This is already implemented for completely different reasons, with stores often exhibiting impulse purchases that they know people struggle to resist in prominent places, and/or around checkouts.

When nudging is moved into the digital landscape, it has the potential for exponential impacts that can transform our social systems because choices in a user interface can be carefully curated, ordered and effectively amplified or promoted. [15] They can also be tailored based on a user's known preferences, previous history on a specific platform, and the Big Five personality traits: openness, conscientiousness, extraversion, agreeableness, and neuroticism; sometimes expressed via the acronym OCEAN. The effectiveness of specific nudges can also be easily calculated, and the approach iterated. The effectiveness of digital nudging for sustainability is still being evaluated. [16] One recent study found that over 60 percent of nudging treatments were statistically significant.

Indeed, digital nudges for promoting sustainable choices are already being used in digital retail channels, from travel and grocery shopping to personal investing. [17] The techniques vary according to the context, but I feel it is important you are made aware of at least six key techniques.

First, using defaults. This involves offering choice conditions in which one of the options is pre-selected. For example, setting a default for offsetting a flight's greenhouse gas emissions when booking a flight. People have to actively deselect buying the offsets or opt-out.

Second, appealing to social norms. This technique emphasizes 'what the majority are doing' and uses its popularity as the 'social-proof' it is the right choice for what 'all' should do. For example, "nine out of ten airline passengers choose the option in their online cart to offset the carbon footprint of their flights" or "for your information, 70% of our previous customers purchased at least three sustainable products".

Third, increasing ease, visibility and convenience. This involves making sustainable options more visible and easier to find. Perhaps placing them first in the search results or using larger images.

Fourth, disclosing more information. This technique makes additional information about the product available, comparable and simple. For example, disclosing the environmental costs or the carbon footprint of an individual product, or across products in the same category. This also involves displaying different certificates that a product has achieved or providing information on the supply chain and product origin.

Fifth, the use of warnings. This approach emphasizes the negative consequences when consumers do not show the intended behavior. It highlights the potential disadvantages of a choice and appeals to the loss aversion tendencies of individuals. For example, displaying a red traffic-light next to a conventional product, and displaying green traffic-lights next to a sustainable product.

Finally, informing people of the nature and consequences of their own past choices. This involves conducting analytics on their historical 'shopping basket', or net level of consumption, to help them understand where their behavior can shift. For example, "Smart disclosure" in the US and the "Midata project" in the UK. This often involves virtuous feedback loops where consumers can see the immediate impact of adopting certain behaviors.

Now imagine these nudging techniques harnessing the collective power of creativity, technology, data, artificial intelligence and other emergent technologies like virtual reality. The result will be the next generation of intelligent influence. Indeed, digital nudges for promoting sustainable choices are already being used in digital retail channels from travel and grocery shopping to personal investing.

Alipay's Ant Forest is the most scaled example of a mobile payment platform that is increasingly enabling and nudging users to shift to greener lifestyles. Ant Forest is an interactive interface embedded in Alipay, the world's largest mobile and online payment platform, where users receive a real-time CO_2 footprint related to many of their everyday behavioral choices. Users gain green energy points as rewards when their CO_2 footprints decrease, and these points can be used to plant virtual trees on the platform, which are eventually translated into real trees. Ant Forest has more than 600 million users who engage with the platform to accumulate virtual green energy points, which in turn have led to the planting of more than 326 million real trees in northwestern China. [18] The success of the platform stems not only from the nudgeing of low-carbon behaviors and the rewarding of energy points, but also from the online social interactions that are supported. These allow players to display their progress, compete with others, share energy, and engage in collective action – a key ingredient in the success of the platform.

Another example of active nudging was developed by a major e-commerce platform. Lazada is Southeast Asia's leading eCommerce platform, backed by Alibaba, with a presence in Indonesia, Malaysia, the Philippines, Singapore, Thailand and Vietnam. The developers of this platform have experimented with developing green consumer segments, in order to shift consumers to more sustainable products. The major conundrum that they are addressing is how to offer green products to consumers who do not usually buy green.

Lazada developed a "Beat Plastic Pollution" collection of products for World Environment Day 2018, and used a combination of marketing, filters and digital nudging to engage people with the green collection. A total of 14 million people engaged in the campaign, while 7 million people purchased products, representing a conversion ratio of 50 percent. This blew all expectations out of the water, demonstrating the directive power of these new digital tools. Imagine if we adopted these technologies, for all the right reasons, at scale!?!

Digital nudging can also be more subtle when combined with other technologies such as gamification, but more on that in Chapter Seven.

Amplifying awareness, agency and collective action

This context presents many unique challenges to the growth of sustainable products and services as well as the authentication of information. We must address these challenges head on by asking: How can trust in various goods, services and information be established in the digital landscape at the speed and scale required to make a positive impact on the planet? For this reason, filtering technologies provide a huge amount of fire power for the force of Socializing.

Filtering is the passing of something through a set of criteria to help narrow options and make a final selection and for this reason has become the essential 'navigation layer' for the ocean of data now at our disposal, whose mechanisms can be divided into two major categories.

On one hand, we have passive filters that are largely user-controlled and simply help us to find and organize information. On the other hand, we have active filters that aim to direct specific actions, choices and behaviors in a manner that is largely driven by algorithms. These often act behind the scenes – with little knowledge, control or informed consent by the user.

Additionally, according to Eli Pariser, author of *The Filter Bubble: How the New Personalized Web Is Changing What We Read and How We Think* [19] what ultimately makes filtering a transformational component for Socializing isn't so much about the what or how something is being filtered but has far more to do with the way in which the end-user comes in contact with filtering. We are either that which is doing the filtering or that which is being filtered.

Our daily lives are influenced by filtering every minute that we spend online as well as increasingly more of the ones we don't.

As a result, none of us are being exposed to impartial choices or unbiased information when we are online anymore. Filtering works behind the scenes of almost every digital interaction, engagement, transaction and outcome we are exposed to. Filtering monitors our search preferences and cross references it with our location, our internet activity, our search history and the plethora of personal data that can be mined from the other platforms also associated with our IP address and discoverable devices.

At each step of the way, a cluster of different filters are creating a personalized avatar of our preferences, likes and dislikes, to increase the level of predictability in our behavior so that it can be influenced for someone else's gain. Our gender, life stage, age, income, interests, family status, political persuasion, religion, culture and more are all being indexed and tracked at global scale.

The algorithms and technologies that underlie our social platforms include perverse incentives that benefit from the spread of information, which in turn, incite strong emotions, such as outrage, rather than truth and cooperation.

What makes this emotional feeding frenzy a serious issue for sustainability is that in most cases our decision-making capabilities on an individual level aren't even being manipulated by the use of factual information. Just because a news story, advertisement or social media post goes viral and incites a strong emotional and behavioral response in its audience doesn't mean the information relayed is based on fact, or in our best interests.

Many people don't fully realize the devastating implications of misinformation and the velocity at which it spreads. In an extensive study carried out by Vosoughi, Roy & Aral, 2018, they found that: *"falsehood diffused significantly farther, faster, deeper, and more broadly than the truth in all categories of information"*. [20] Another recent study found that false news stories are 70 percent more likely to be retweeted than true news stories are. It also takes true stories about six times as long to reach 1,500 people, as it does for false stories to reach the same number of people. [21]

Indeed, it is deeply concerning that when we are exposed to fake news and misinformation about climate change two outcomes can occur. We either end up discounting both the fake and the facts or confirmation bias kicks in and we only believe what we already believe to be true. [22] Climate deniers discount and discard scientific information about climate change but celebrate false claims about climate change as these reinforce their world view.

We also become less aware of the weight of expert opinion backing true information. As we are exposed to increasing levels of misinformation, we become less convinced that scientists have reached a nearly universal consensus on climate change. This often leads to polarization between different groups on the level of scientific certainty and on the magnitude of the existential crisis we face. This division undermines collective action and prevents climate change from becoming a unifying political focus. [23] This is known as the 'consensus gap' and causes "truth decay". [24] This puts us at significant risk of compromising the credibility of the fact-based information and scientific evidence we need to rely upon to put our planet right. [25]

The real McCoy

The ultimate impact of fake news and misinformation is that it normalizes indifference to truth and undermines trust in our public institutions. When people do not share a common baseline of facts, it becomes impossible to have civil discourse and rational debate that is essential to a functioning democracy – and even more impossible to catalyze collective action. [26]

Social media platforms have been accused of amplifying filter bubbles in the US for example – leading to completely different views of our present reality, and deep political polarization. [27]

News and information about our climate crisis, as well as awareness of and social proof of sustainability solutions, is not being amplified across social groups at the same rate as some other forms of perhaps more sensational content, because it does not trigger the same level of collective emotional response and is therefore not considered 'newsworthy' or 'sticky'.

Our tendency to get caught in filter bubbles and echo chambers can be addressed through algorithms that detect this tendency and start to feed us alternative perspectives. This will become increasingly important as fake news is now proliferating in these echo chambers, and then causing real-world impacts. Our need to reshape facts and make decisions in a manner that makes us feel secure and in control can be combated by technology that can verify and track the provenance and authenticity of information.

After the Cambridge Analytica/Facebook scandal blew up, exposing many of the shadows of this force, social media has been under more scrutiny than ever. Twitter, Facebook, Youtube and others are responding with varying degrees of self-imposed standards. The Filter Bubble Transparency Act (FBTA) has even been passed through the US Congress [28] – but we need to ask – is it enough? While new standards are being developed on these platforms, they focus primarily on political and racial tensions and less on how misinformation is impacting social action on climate change or sustainability.

So, how do we make news and information on climate change and sustainability ring true again?

Globally, with respect to climate change, there are many initiatives working with authenticating technologies, to fight fake news and restore public trust in both the content we read and the sources that content comes from. [29]

Climate Feedback is one such organization and uses a web-based content annotation tool that relies on crowdsourcing to engage a worldwide network of scientists to sort fact from fiction in climate change media coverage. Their goal is to "help readers know which news to trust". The process results in a credibility score for the articles and photos they review.

Bad News is another great tool for educators and members of the public that is intended to build user understanding of the techniques involved in the dissemination of disinformation. This game exposes players to fake news tactics that are commonly used against them online by putting them in the pseudo position of a news baron for fake news. Players win by publishing headlines that attract the most followers.

One of the most awarded organizations using filtering as a force for good, to combat misinformation, is a platform known as Factmata. According to their website, Factmata develops contextual understanding algorithms, driven by communities and experts who share the goal of reducing online misinformation and abusive content from the internet. It uses unique algorithms combined with AI to classify content in subtle ways, building a platform to help fix the problem across the whole of the media industry – from the spread of biased, incorrect content and misleading clickbait on numerous aggregating platforms, to the use of advertising networks that help distribute that content. The team has won many awards, including the UNESCO Netexplo Award 2019.

Alto Analytics is a Madrid-based big data analytics start-up that is bridging the gap between awareness, buying behaviors and decisioning. They are doing this by mapping digital relationships and patterns, in order to fight disinformation and deep fakes, guard a brand's reputation and provide commercial insights and online/offline influence analysis. With the capability to do real-time analysis of public data sources across 53 languages and 125 countries, they could be complete game changers inasmuch as how we choose to interact with available information as well as allow organizations to interact with our personal data-trail. [30]

From data to action

One of the challenges faced by the sustainability movement has been the translation of data about the planetary crisis into emotional stories and experiences that change behaviors and social norms. Indeed, the key to any behavioral change is telling stories that touch human emotions, generate empathy, and become reflected in social norms. Facts don't always change our minds – but peer pressure and social proof metrics often do.

That is why it is critically important to find ways for the digital world to speak to human emotions in the physical world. "*Data without emotion doesn't bring about change*". If we want to drive change using data, we need to bring in emotions, empathy, narratives and personal experiences. [31] We need to go beyond the 1s and 0s of the digital landscape – and keep in mind the human experience. We must connect the logical and emotional centers in our brain in a manner that triggers a reaction, action and behavioral shift.

Immersive technologies use a combination of digital technologies including AI, dynamic data visualization, virtual reality and extended reality to achieve this goal through the creation of fully immersive experiences. This is a process that takes a user from observing (through a window) to digitally immersing as many senses as possible in a curated experience, in the process engaging the user in content that creates the feeling of a real lived experience or, at the very least, an extremely believable illusion of one. Immersive technologies are powerful enough to invoke a memory, create a felt sense of empathy, as well as other emotions, and/or a strong human connection.

It is also important to cite the differences between engagement and immersion as they are not even close to the same thing. Engagement takes place when a story, or a marketing message, provokes some sort of action among the audience – a tweet, a post, a zoom call, a purchase. Engagement is the desired result of most tech platforms as described above. Immersion on the other hand takes place when the audience forgets that it's an audience at all. Immersion blurs the lines – between story and marketing, storyteller and audience, illusion and reality. That is what gives it the ability to make such an enormous impact.

Steven Spielberg, talking on stage with George Lucas at the University of California famously stated, "*We're never going to be totally immersive as long as we're looking at a square, We need to get rid of that and put the viewer inside the experience, where no matter where you look, you're surrounded by a three-dimensional experience. That's the future*". And there is much evidence to suggest that digital immersion is leading to important behavioral outcomes. For example, Facebook claims that 48 percent of viewers that experienced charity content using virtual reality were likely to donate to the causes that they experienced. [32] A similar report by the United Nations noted that their

VR production Clouds over Sidra, which films the life of a twelve-year-old Syrian refugee, "helped raise twice the charity's normal rate". [33]

Fortunately a wave of VR applications aimed at environmental sustainability is emerging. Of course, the ultimate prize in the VR landscape is helping people experiment with sustainability challenges at a planetary scale, and then helping them to understand and adopt specific behaviors and actions they have the personal capacity to execute upon. These are monitored and reinforced by mobile apps and algorithms, which assist the effectiveness of the process.

The breathtaking work of TeamLab Art from Japan offers a fully immersive experience within a three-dimensional digital art space. Viewers, in interaction with their environment, can instigate perpetual change in an artwork. Through an interactive relationship between the viewers and the artwork, viewers become an intrinsic part of that piece. TeamLab believes that the digital domain can expand the capacities of art, and that digital art itself can create new relationships between people, as well as transformational emotional responses. This kind of technology could be fundamental in the sustainable toolkit, for immersing people in the challenges faced by the planet and helping them understand the connections between their individual behaviors and planetary health.

The practice of world building – popularized by games such as Sim City – also offers important opportunities to demonstrate the potential impact of different environmental and climate scenarios. [34]

The World Building Institute is a cutting-edge USC nonprofit Organized Research Unit dedicated to the dissemination, education, and appreciation of the future of narrative media through World Building. The term 'World Building' designates a narrative practice in which the design of a world precedes the telling of a story; the richly detailed world becomes a container for narrative, producing stories that emerge logically and organically from its well-designed core.

There are three core beliefs at the heart of 'World Building', namely that storytelling is the most powerful system for the advancement of human capability due to its ability to allow the human imagination to precede the realization of thought; that all stories emerge logically and intuitively from the worlds that create them; and that new technologies powerfully enable us to sculpt the imagination into existence.

One of the reasons why virtual reality technology is so powerful is that it offers full immersion and focused attention by the user. It prevents a user from being distracted by the barrage of gadgets associated with the attention economy and offers them an extended opportunity to focus on a specific issue. It can help users visualize and experience the implications of global environmental degradation and climate change, as well as inspire them by illustrating what a sustainable future looks like.

But augmented reality is also demonstrating the value it can have in terms of raising awareness and catalyzing action. Here the world of the Mila

Institute in Montreal, Canada is also trendsetting. Their work aims to raise awareness and conceptual understanding of climate change by bringing the future closer and enabling users to experience the impacts. This is achieved by depicting accurate and personalized outcomes of climate change in their homes and neighborhoods, using cutting-edge techniques from artificial intelligence and climate modeling.

The institute offers an educational tool that produces accurate and vivid renderings of the future outcomes of climate change, which depicts the way that such trends are likely to affect individuals, in order to motivate behavioral change. Other augmented reality applications for visualizing environmental change include "This is Climate Change", "We Live in an Ocean of Air", "Greenland Melting", "Immerse", "Osmose" and "After Ice".

The hope in using these tools is to catalyze a perceptual shift to a global mindset – a wider view of ourselves and our role in a global community that can help to spark global-scale empathy. A further intention is to kick-start the transformation from 'me' to 'we', and to reinforce the link between education, engagement and conscious sustainability.

Summary

As the only way we achieve the outcome of living on a sustainable planet is based on activating a whole world transformation, it stands to reason that the global collective will be made up of an extremely diverse human elements including all races, cultures, ages, backgrounds, professions, education levels etc. Their interaction and engagement with each other and among their stakeholder groups makes up the most valuable resource sustainability has in its corner and is the core driver and organizer behind the success of it.

Given the vital role the global collective will play and the diversity of not only each individual person, their countries and organizations in representation but also the roles each person plays, we need to assume that a cookie cutter approach to engaging on a whole, with the populations that are part of our ecosystem, won't suffice. This is where the transformational force of Socializing plays a critical role in supporting successful sustainability outcomes and, when fully harnessed, can, first, help subscribe entire populations to a more sustainability focused agenda and lifestyle, at speed and scale.

Second, we can be connected to forms of hive intelligence to learn from others who have tackled similar situations in the past, ensuring that we face our reality head-on, and take an effective fact-based course of action. We can also design algorithms to help amplify virtues and sustainability solutions, rather than vices and outrage. We can project the social proof of sustainability initiatives, behaviors and solutions onto the nearly four billion people active on social media and other digital platforms in a manner that builds confidence and trust, as well as underpinning collaboration and collective action. The transformational force of Socializing can amplify the social proof of sustainability initiatives to evoke the emotional responses needed to help us take ownership of sustainability outcomes and catalyze the necessary changes in our individual behavior. By Socializing information about initiatives that are making a massive difference, groups and populations are motivated and inspired to join in efforts and conversely, by being able to connect with the plight of less successful initiatives, the call for collective action is more likely to be answered.

Third, it can be used to detect and manage a range of cognitive biases as well as help us connect with that which is outside of our immediate experience, empowering us to make better choices and more considered decisions than we might have made otherwise.

And yet to achieve these transformative outcomes, many enablers are still needed.

We need to deeply contemplate the human element in all of this. How will we as humans respond to and adopt the sustainability data, insights, tools, platforms, products and services that are developed? We can try to pump as many facts as possible about the existential risks that we face into people's minds, but this won't necessarily change their opinions, let alone motivate action. That is why implementing real-time feedback on how new incentive structures are influencing consumers into the adoption strategy of new products, services, behaviors and lifestyle choices is critical.

Yet the above enabling condition is fraught with opportunity for manipulation through bad-faith discourse and perverse incentives. To prevent such things, we also need more rigorous standards and safeguards to prevent the incorporation of misinformation and fake news into algorithms and platforms. The exposure to misinformation and fake news can have devastating consequences for both people and sustainability-focused applications, in terms of altering perceptions, motivations, and behaviors, including buying decisions, and more. This can severely undermine positive action, preventing the amplification of sustainability impacts at the systems level.

References

1. Kemp, S. (2020). Digital 2020: 3.8 Billion People Use Social Media. Available at https://wearesocial.com. Accessed January 30.
2. World Economic Forum. (2022). *Shaping the Future of Digital Economy and New Value Creation.*
3. Oberlo. (2022). *How Many People Have Smartphones in 2022?* Available at www.au.oberlo.com
4. Kemp, S. (2020). *Digital 2020: 3.8 Billion People Use Social Media.* Available at https://wearesocial.com. Accessed January 30.
5. Ibid.
6. Ibid.
7. Dredge, S. (2015). "Zuckerberg: one in seven people on the planet used Facebook on Monday." *Guardian.* Available at www.theguardian.com. Accessed August 27.

8. Kemp, S. (2020). *Digital 2020: 3.8 Billion People Use Social Media*. Available at https://wearesocial.com.
9. Oberlo. (2022). *How Many People Have Smartphones in 2022?*
10. Kemp, S. (2020). *Digital 2020: 3.8 Billion People Use Social Media*. Available at https://wearesocial.com.
11. Elias, J. & Petrova, M. (2019). *Google's Rocky Path to Email Domination*. CNBC.
12. Kristensson, P. et al. (2017). "Influencing consumers to choose environment friendly offerings: Evidence from field experiments." *Journal of Business Research* 76: 89–97.
13. Vermeir, I. & Verbeke, W. (2006). "Sustainable food consumption: exploring the consumer 'attitude – behavioral intention' gap." *Journal of Agricultural and Environmental Ethics* 19: 169–194.
14. Thaler, R. & Sunstein, C. (2009). *Nudge: Improving Decisions About Health, Wealth and Happiness*. Penguin.
15. Dhar, J. et al. (2017). *The Persuasive Power of the Digital Nudge*. BCG.
16. Hummel, D. & Maedche, A. (2019). "How effective is nudging? A quantitative review on the effect sizes and limits of empirical nudging studies." *Journal of Behavioral and Experimental Economics* 80, March.
17. Albrecht, M. (2020). "More "nudging" for pro-environmental behavior." *Journal of Behavior Studies in Organization*, 3: 13–21.
18. Green Digital Finance Alliance. (2020). *Fintech for Biodiversity: A Global Landscape*. Available at www.naturefinance.net
19. Pariser, E. (2011). *The Filter Bubble: How the New Personalized Web Is Changing What We Read and How We Think*. Penguin.
20. Vosoughi, S. et al. (2018). "The spread of true and false news online." *Science* 359 (6380): 1146–1151.
21. Dizikes, P. (2018). *Study: On Twitter, False News Travels Faster than True Stories*. Massachusetts Institute of Technology.
22. Smith, J. (2017). *Fake News Allegations vs. the Facts – How it Affects Sustainability*. Available at www.eco-officegals.com, Accessed May 7.
23. Reid, L. (2020). *Fake News Week 2020: Exploring the Shocking Scale of Climate Change Misinformation*. Available at www.brandwatch.com. Accessed March 30.
24. Cook, J. (2019). Understanding and Countering Misinformation about Climate Change, in *Handbook of Research on Deception, Fake News, and Misinformation Online*, eds. Chiluwa, I. & Samoilenko, pp. 281–306. IGI-Global.
25. Wilding.D., et al. (2018). *The Impact of Digital Platforms on News and Journalistic Content*. University of Technology Sydney, NSW. Centre for Media Transition
26. Kakutani, M. (2018). *The Death of Truth*. William Collins.
27. Molla, R. (2020). *Social Media is Making a Bad Political Situation Worse*. Available at www.vox.com. Accessed November 10.
28. US Congress. (2021). *S.2024 – Filter Bubble Transparency Act*. Available at www.congress.gov/bill/117th-congress/senate-bill/2024
29. RAND Corporation. (2022). *Tools That Fight Disinformation Online*. Available at www.rand.org
30. Alto Analytics. (2022). *Introducing Constella, the Leader in Digital Risk Protection*. Available at www.alto-analytics.com. Accessed December 15.
31. Cotgreave, A. (2015). *Data Without Emotion Doesn't Bring aboutC*. Availale at www.infoworld.com. Accessed October 6.

32. Samit, J. (2017). *How These Charities Are Using Virtual Reality to Reach Donors This Holiday Season.* Available at www.fortune.com. Accessed November 22.
33. Nelson, K. et al. (2020). *Virtual Reality as a Tool for Environmental Conservation and Fundraising.* PLoS ONE. 15 (4)
34. Evans, A. (2021). *On Virtual Worlds.* Available at www.global dashboard.org. Accessed January 13.

6 Transformational Force #3
Valuing

Defining the force

Valuing is the process of quantifying and calculating the contribution that a product or service is making to the economy, to human well-being, and to planetary sustainability. Valuing translates our natural world, as well as the costs associated with our impact on it, so that it can be integrated into our economy.

Ultimately, the digital technologies that form the backbone of this force can drive outcomes, such as sustainability, circularity and regeneration. These outcomes are supported by the digital technologies of Valuing in three distinct ways:

First, through **tracking**, **tracing** and **quantifying** inputs and outputs, such as natural capital, ecosystem services and carbon emissions, in order to **translate** these costs into accounting practices and business models. This also creates full supply chain **transparency** and **accountability** by enabling the **disclosure** of this information in digital formats to different actors.

Second, through changing the way that we can **express** and **exchange** a complex range of values, beyond price and profit, in the marketplace. This includes **matchmaking** consumer values with the environmental and social performance of products and services.

Third, Valuing **monetizes** certain ecosystem services or environmental solutions into economic parameters, enabling **market exchange and value sharing** across stakeholders.

The rise of multi-dimensional digital value

I think that anyone reading this book would agree that if we are to truly achieve a circular economy and a sustainable future for our species and planet, we must begin to define and measure our economic success in more ways than exclusively focusing on maximizing profit. [1]

That is why I have identified Valuing as a transformational force – because it can play an essential role in sustainability across the three levels of digital transformation, as they relate to the economy. Recall:

DOI: 10.4324/9781003187523-8

Level 1 (Products/Services) Drives supply-chain transparency in all inputs, outputs and externalities of a product or service.

Level 2 (Brakes and Accelerators) Applies full-cost accounting to internalizing environment and climate externalities into the business model.

Level 3 (Digital Ecosystems) Develops business models and financial instruments that translate social value and environmental impact into incentives and behavioral change at systems level.

As these levels suggest, Valuing provides a range of associated variables that now enable us to express and transact across a much broader range of our values – beyond price and profit – and in more efficient and diverse ways. Origin, provenance, environmental performance, carbon performance, durability, reusability, circularity, social impact, respect for human rights, labor standards, fair wages, and many other factors can now all be captured and, in doing so, can now be translated into balance sheets. Valuing can help account for environmental externalities in value chains and bring transparency to the social good companies are generating. It can establish the foundation for a circular economy by tracking and tracing material flows. It can help monetize ecosystem services and help distribute payments for their protection. It can also create new business models that tackle the question of advertising and subsidies that perpetuate unsustainable behaviors.

As the economy digitalizes, algorithms and code are starting to underpin and mediate all economic activity and transactions, as well as much of our social relationships, education, health and lifestyles.

If we want to have any chance to achieve the SDGs, stop climate change, protect biodiversity, and prevent pollution in the next ten years, we need to focus our attention on influencing these algorithms and codes to ensure they are optimizing not only for profit, but also for people and planet. This is where the true transformational opportunity for Valuing as a force for good lies.

Today, tens of millions of businesses depend on digital platforms and digital marketplaces. An estimated 1.9 billion people worldwide are purchasing goods online, amounting to USD3.5 trillion of sales in 2019. In 2022, the World Economic Forum (WEF) estimated that 60 percent of global GDP was digitalized, and that over 70 percent of new value created in the economy over the next decade will be based on digitally enabled platforms. [2]

This is transformative, in terms of us finally being able to address many of the core barriers to sustainability within our Economic Operating System, as well as how this system drives, interacts and interconnects with other systems.

Let's have a look at how the transformational force of Valuing can weave itself into our economy to achieve our sustainability goals.

Tracking, tracing and translating the new metrics of value

The first major contribution that the force of valuing can make to speeding and scaling sustainability is to effectively track raw materials, and the products they are made into, through every step of their life cycle. This then enables us to effectively trace the environmental impact of these materials and products during every stage of their lifecycle.

Valuing can, therefore, help us internalize the environmental costs of extraction, production, distribution, consumption and disposal. It does this by enabling the translation of the intangible costs associated with these environmental impacts, so they can then be attributed toward the tangible production cost of a product. This helps to ensure that the environmental impact caused by conducting business as usual is as important on the balance sheet as any other factor associated with the production of goods and services.

Thankfully, there is now a range of digital technologies enabling this process of tracking, tracing and translating.

GS1 is a not-for-profit, intergovernmental organization that is responsible for developing global standards for supply chain management. The GS1 international barcode is poised to become a major enabler to track value transformation along the value chain. [3] GS1 is developing global and open standards to underpin the Circular Economy concept, in order to offer an enhanced level of transparency and traceability regarding specific products. [4] This includes product packaging, the provenance of raw materials, the conditions of production, and the identification of chemical substances.

GS1 in Italy is pilot testing a "Barcode for Environment". The aim of the project is to effectively share environmental attributes using GS1 standards with all partners along the value chain. The intention is to make information on the environmental footprint clear, understandable and comparable for all consumers.

QR codes are another important digital technology that can store detailed data about a product, enabling immediate retrieval by consumers, as well as processing by filters and recommendation algorithms.

Many products and stores are starting to deploy this technology at scale. In Alibaba's new Hema grocery store chain, consumers can scan QR codes for every product on sale, in the process learning about the origin of products, along with provenance, environmental and social certifications, food miles, nutritional information, allergy risks, recipes and other information. [5] The Hema app can also make personalized product recommendations, helping consumers to identify and locate ethical products.

The concept of a digital product passport is also being proposed as part of the European Green Deal. Digital 'product passports' will be developed for products that have potential to enter the circular economy and therefore extend their lifecycle beyond being a single use item. The passport would enable the tracking and tracing of products, while also containing information on origin, composition (including the presence of substances of concern), critical raw materials content, recycled material content, possibilities or instructions for recycling, re-use, repair and collection upon discarding, dismantling, and handling as waste, among other characteristics.

Every product would receive a unique identifier (think 'birth certificate') with basic information (producer, model, date); this is kept on a centralized registry. An address is created like a URL (Uniform Resource Locator) for the product. When combined with a tag (QR code, RFID, Bluetooth tag), the company, consumer or public authority can connect directly to access the product's unique digital profile, with quantitative and qualitative, static and dynamic, standardized and machine-readable data. Most of the data stream will remain in the place of origin, and only connect into the product passport by API. The registries with links to the distributed data can then be managed by trusted third-party partners – ensuring reduced cost and/or administrative burden.

The information contained in the digital product passport would play a huge role in assisting with the automation of purchasing decisions by consumers, and especially by eCommerce algorithms that are optimizing for specific sustainability criteria. It would also be transformative in tracking the life story of a product, enabling services related to its remanufacturing, reparability, second-life, recyclability and new business models.

Of course, one of the killer technology applications on the horizon for supply chain tracking is blockchain. Yes, you knew we had to bring up blockchain again at some point! While it has gone through rounds of hope and hype, it really is a groundbreaking technology.

At its heart, blockchain is simply a form of digital record keeping, but one that is vastly superior to the analog version, as digital records stored in the blockchain are both permanent and tamper-proof. This immutable quality of the blockchain is often asserted as being one of its major qualities. Its ledger of digital records is achieved using a distributed database that deploys cryptography to authenticate digital assets and transactions. Blockchain technologies help solve fundamental problems linked to data-sovereignty, provenance and ownership by supporting authenticated, trustworthy, and verified transactions between parties at a global scale.

Everledger is a brilliant example of blockchain technology, developed to create a secure and permanent digital record of an asset's origin, characteristics and ownership. The company's blockchain technology[6] is now being applied to the full lifecycle of lithium-ion batteries, as part of their support of the Global Battery Alliance. [7] This transparency helps industries respond to growing expectations for sustainable, verifiable sourcing, and assists consumers to participate in the process of extending the lifecycle of their products.

Driving transparency and accountability through disclosing environmental and climate performance

As well as unleashing a new era of measuring environmental impact and performance, digital technologies are also driving greater transparency across supply chains. Companies are finding it increasingly difficult to ignore or hide environmental issues related to their supply chains and are under major pressure to improve their environmental performance. Many consumers and investors are also seeking to support companies with strong environmental and climate track records.

A recent survey conducted by the Commonwealth Bank of Australia forecast that there will be a 73 percent increase in the demand for such products and services within the next three years, [8] and that this demand will become exponential within a five to ten years' timeframe. In fact, according to the World Economic Forum, 72 percent of people between the ages of 15 and 20 are willing to pay extra for environmentally and socially-responsible products and services. [9]

It should come as no surprise then that companies with a strong track record for aligning their business outcomes with environmental performance are beginning to communicate this position to their consumer base as part of their marketing, which has become known as communicating their Sustainability Value Proposition (SVP). Often the motivation for doing this is to gain an edge on their competitors. For consumers – green has finally become the new black!

However, historically, from an investor's perspective, there has been a complete disconnection between the market value of a company and the positive or negative contribution it makes to society and the planet. The metric of price, in terms of a company's stock market valuation, has created a significant barrier with respect to understanding the overall social or environmental value that a company is contributing to the world. The incessant focus on quarterly stock price and financial returns to shareholders forces companies into a short-term time horizon. The 'non-financial' or 'extra-financial' risk and performance linked to issues such as climate change exposure is difficult to assess or compare across companies, supply chains or sectors. Furthermore, there is often an incentive to conceal information that could negatively impact stock prices.

In many cases, transparency is antithetical to companies, as it reduces informational asymmetries that can be used to hide actual risk and value. The fact that our pensions and retirement benefits are often based on stock market returns is another perverse incentive that prevents systemic change. This creates a kind of fundamental lock-in to stock market returns at the individual level – even when people know it is not sustainable at the civilizational level.

It is imperative that we shift away from the 'infinite growth' mindset, with its focus on maximizing shareholder value. Instead, it would be preferable to define and communicate how such growth contributes to the overall public good, in the process giving a holistic and, in my opinion, more realistic market valuation.

Initiatives such as the G20 Task Force on Climate-related Financial Disclosures (TCFD) are also critical frameworks for connecting financial risk with climate risks that speak to the above problem. By requiring companies to disclose specific information, the TCFD reporting recommendations enable stakeholders to better understand the concentrations of carbon-related assets in the financial sector, and the financial system's exposures to climate-related risks.

The more climate risk, the more it affects their valuation on the share-market.

Positive environmental performance is now being recognized as an economic imperative. As such, companies are now scrambling to showcase their commitment to environmental sustainability in a visible and verified manner. As a result, we have seen the rapid development of corporate and sovereign Environment, Social and Governance (ESG) frameworks.

As discussed in Chapter Four, ESG frameworks largely focus on mitigating impacts and improving the efficiency of resource use across supply chains. When it comes to translating ESG frameworks into share market value, we are making progress, albeit slowly. Of the USD95 trillion invested in global stock markets, only USD2 trillion is aligned to Environment, Social and Governance (ESG) principles. [10] So we need to ensure that emerging fintech applications begin to hardcode sustainability considerations, so that they can be used to mobilize finance for sustainability.

A potential breakthrough in helping to objectively apply value to the ESG performance of companies is provided by the emerging field of 'spatial finance'. At the core of spatial finance lies remote sensing and geo-spatial data collected from IoT sensors. New generations of small satellites are orbiting our planet, taking high resolution images of every point on Earth, enabling us to observe planetary-scale change on a daily, weekly or monthly basis. For example, companies such as Bluefield incorporate data from 23 different satellites, a range of ground sensors (such as airport anemometers), government records, and other reliable independent data sources in order to detect, pinpoint, and quantify greenhouse gas emitters around the world on a daily basis.

When combined with artificial intelligence (AI) to automatically scan and interpret this vast amount of visual data, the true transformational capabilities of spatial finance become apparent. This innovation can provide near real-time information on how environmental and climate risks, opportunities, and impacts are managed across a supply chain. Spatial finance also enables financial markets to more accurately measure and manage sustainability-related risks, as well as enhancing a vast range of other factors that affect risk and return in different asset classes. The development of spatial finance furthermore provides insights at differing scales for different applications, from the asset-scale for project finance through company-scale for investment, to country-scale for sovereign debt. A range of companies are now incorporating these tools into their analytical services, including Refinitiv, Bloomberg, S&P Global, and Reprisk.

As consumers, investors, governments and companies continue to further align their consumption goals with sustainability values, market incentives will shift. There will be a massive divestment out of 'toxic assets' that have unsustainable footprints. This also paves the way for financial investors to have a real-time view on the concentrations of carbon-related assets in their portfolios, together with exposures to environment and climate-related risks. This will instigate an even greater shift towards ESG frameworks.

However, some companies go further than this ESG performance, wishing to also communicate the wider positive purpose and impact that they hope to make. Often, such aims are couched in terms of contributing to specific SDG goals. This also poses two challenges in terms of valuing and communicating this intent. First, how to provide trustworthy calculations that prove a company has taken action on its promises, and, second, how to disclose this into actionable information to inform decision-making by consumers.

These twin challenges are being addressed by Digital With Purpose (DWP) – an international movement by the Global Enabling Sustainability Initiative (GESI), [11] which aims to fundamentally change the business models of digital companies by measuring how their products and services can contribute positively to SDGs. Digital With Purpose is an example of the newly coined term 'corporate digital responsibility', which summarizes the emerging responsibilities of corporations regarding their digitalization-related impacts, risks, challenges and opportunities. [12] DWP is creating a series of metrics, a certification scheme, and a 'digital stamp' that will help companies measure and communicate how their digital products, services and business practices are, not only achieving ESG targets, but also contributing to mission-oriented positive outcomes for social and environmental change. This initiative can help consumers, ranging from individual decision-makers to companies and governments, more easily identify products that match their priorities and values. The initiative is being supported by Deloitte to ensure the framework is robust, auditable and trustworthy.

Matchmaking values and automating procurement with purpose

Once performance information and detailed metrics are publicly disclosed, the next challenge that valuing can overcome is matchmaking. In this process, algorithms are needed to match a complex range of consumer or investor preferences and values with the detailed performance metrics of products and services. We are increasingly moving into an era of procuring with purpose. Consumers and investors, including individuals, financial institutions, companies and governments, now, more than ever before, feel the responsibility for ensuring that their dollars make positive social and environmental impacts.

Public and private procurement of goods and services is a major driver of the economy, and a fundamental leverage point to scale environmental sustainability. Public procurement by governments accounts for 15–20 percent of global GDP, [13] while direct business-to-business (B2B) transactions on eCommerce platforms alone represent USD21 trillion, [14] or around 23 percent of global GDP. Surprisingly, B2B eCommerce transactions are five times larger than business-to-consumer (B2C) eCommerce transactions. [15]

One of the companies that is pioneering the digitalization of procurement with purpose through their supply chains is the Ariba Network of SAP. Ariba is a global network where businesses connect, communicate and collaborate. It is the largest business-to-business (B2B) network in the world, connecting more than 4.4 million companies in 190 countries, and facilitating more than USD2.9 trillion worth of transactions annually. [16] On the Ariba Network, companies have a single, centralized marketplace solely focused on the sustainability agenda to identify opportunities, find innovative solutions, securely share information, and close deals.

SAP Ariba has fully embraced the idea of procuring with purpose and is helping companies adopt sustainability and transparency as a competitive strategy. SAP offers full transparency and visibility across a company's supply chain to better assess the real-time social, economic and environmental impact of transactions.

Ariba offers a value lifecycle manager tool to help companies assess and benchmark their goals and key performance indicators (KPIs). This helps clients prioritize problem areas and compare performance with their peers against 100 KPIs.

But Ariba's social impact work doesn't stop there. [17] As Ariba has also embraced the concept of a digital ecosystem, companies on the Ariba Network can also now seamlessly connect to social good digital platforms to directly fund nonprofit programs aligned with their corporate sustainability goals. [18]

Traditional financial players, such as Mastercard, are also getting into the sustainability matchmaking game. In 2021, Mastercard and 'Helpful' partnered to launch a free consumer debit card to empower sustainable living. The HELPFUL debit Mastercard offers a financial solution that puts the planet first, appealing to a new breed of conscious consumers who want their purchases to reflect their values, principles and lifestyles.

With over 150 sustainable brands on board, the partnership establishes one of the UK's largest sustainable retail directories. Consumers who shop with brands from the directory will also receive a reward of up to 10 percent cashback. HELPFUL further pledges to plant a tree for every purchase made on their cards. Mastercard is another company leading the way in terms of carbon valuing, by being the first credit card company to offer a carbon calculator to their clients. Individual carbon footprints are tracked month-by-month across a variety of spending categories, so that users can better understand where they are having the greatest impact. [19]

CoGo – short for "connecting good" – is a free mobile app for ethical consumption. Users can choose from 13 social and environmental values, and the app helps them discover third-party accredited businesses that align with their values. Users simply identify the values that are most important to them, whether that's vegan brands, social enterprises, or companies that are carbon neutral or positive. It's also possible to find businesses that are making an impact through the app, search for independent and sustainable businesses in the immediate vicinity of a user, and also those companies that are in line with particular ethical values.

CoGo is one of the first tech innovations of its kind to aim at changing the way that consumers spend their money, while encouraging them to connect with businesses that share the same values about the planet. CoGo also calculates your personalized carbon footprint in real-time, linked directly to your spending transactions and lifestyle choices.

The transformational force of Valuing hasn't just opened the door for consumers to connect more easily with products and brands that align with their values, it has also enabled entirely new opportunities for the monetization of environmental goods and services to come into existence.

Monetizing environmental goods and services

One of the huge benefits of digital technologies is the explosion of new opportunities for the monetization of environmental goods, services, and values that were not previously captured by economic models and financial instruments.

In some cases, our economic operating system failed to account for environmental impacts precisely because environmental goods and services had no economic value. Here too, the digital technologies that power the force of Valuing are helping to measure and monetize environmental goods and services, so they can be more readily exchanged in a global digital marketplace.

In this way Valuing doesn't just connect consumers to products and companies that are more sustainable, but, arguably more critically, it also enables the connection of companies to finance. And technology's fintech vertical is set to play a significant role in that process.

Fintech is the term used to describe the integration of technology into financial services organizations, while "Green Fintech" is the term used to describe fintech-focused organizations that have put sustainable outcomes at the core of their philosophy and business offering. The global fintech market is projected to reach almost USD500 billion by 2025, and Green Fintech is set to represent an increasing share of this fast-growing space. [20] Blackrock Capital, one of the most successful investment and risk management firms in the world, and a firm also recognized for their support of the greening of fintech, supported 75 percent of climate and social proposals pitched to them in the first quarter of 2021. [21] Each year Blackrock's CEO, Larry Fink, writes a letter to their clients stressing more than the last the importance of 'greening' financial services. Fink observes: *"as markets start to price climate risk into the value of securities, it will spark a fundamental reallocation of capital"*.

Green bonds are one such example of this phenomenon. Green bonds are a growing green fintech category, comprising fixed-income securities that raise capital for projects with environmental benefits, such as renewable energy or low-carbon transport. [22] Although they make up a small fraction of the overall debt market, green bonds are attracting attention because the need to meet net-zero emissions targets will require trillions of dollars of capital invested in them from both the public and private sectors.

By the end of 2022, the annual issuance of green bonds reached a record high of USD395.5 billion and, according to the Climate Bonds Initiative, is tipped to reach an annual issuance of 1 trillion by end of 2023 and exceed a cumulative issuance of USD25 trillion by the end of 2025.

One of the challenges faced by green bonds is that they require an additional layer of 'green performance' data and 'proof of impact'. The issuers of green bonds must have meaningful data to report to the bondholders about how the underlying asset (e.g., wind turbine, solar panel, and so forth) is performing.

Fortunately, blockchain technology collaborating with both IoT and AI might offer a solution. Sensors can be installed into the hardware of the green asset, which monitor usage and measure the corresponding reduction in carbon usage, but then also upload that data directly to the blockchain. Once there, AI can help interpret the data, as it reaches the investor. This points towards a future where one of the biggest constraints to the growth of the green bond market – the difficulty and expense of reporting on impact – is lifted. [23]

Using digital technologies to automate the measurement, monitoring and monetization of environmental assets is also set to revolutionize the renewable energy sector. Renewable Energy Certificates (RECs) are widely used market instruments that enable corporate purchasers of renewable energy to make third-party verifiable claims about the sources of their electricity. RECs, GOs, TIGRs, and other such electricity market instruments have become powerful catalysts, driving investment into renewables in many markets. These instruments enable organizations to ensure that their electricity is provided from renewable sources, which produce low- or zero-emissions, thereby reducing the organization's market-based scope 2 emissions.

One of the challenges faced by RECs is the inability to monitor and verify highly distributed sources of renewable energy production. As a result, REC markets have largely been established in mature markets, with utility-scale renewable energy deployments that are centralized and trusted. However, this has constrained investments in developing countries, where investments in distributed renewable energy could have a larger GHG emissions reduction impact. While 35 percent of renewable capacity growth through 2050 will be distributed, RECs to date have been unable to capture this value.

Digital technologies are now helping to remove this investment barrier. Real-time digital monitoring technologies will be used to certify and value distributed renewable electricity attributes. These will be monetized through the creation of a new, internationally recognized market instrument referred to as the D-REC (Distributed Renewable Energy Certificate).

The approach is straightforward. The process begins when a project developer installs a distributed renewable energy system, such as a solar home system, commercial rooftop system, community mini-grid, or campus microgrid. As electricity is generated, data from the renewable system is transmitted to the D-REC monitoring and tracking platform. Generation data from multiple renewable systems are aggregated to create a D-REC in accordance with the protocols of international standards organizations. The corporate buyer pays a project aggregator for the D-RECs generated from the underlying projects, and proceeds of the sale flow back to the project developer. Digital technologies are helping to capture value, as well as share those benefits with both D-REC buyers and producers. Ideally, additional capital is attracted to the project because of the improved economics, and distribution of benefits, that are, in many cases, capable of being community wide.

However, while these kinds of tech-savvy solutions may work well for environmental investments that have a digital backbone, it might not be as suitable for nature-based solutions. Presently, only 10 percent of all green-bond proceeds were allocated to nature – a clear indication that biodiversity assets are not yet sufficiently 'wired' to reach capital markets. [24] The gap between people and nature in the current engagement model for biodiversity finance no doubt partly explains the corresponding gap in funding: a shortfall of USD300–400 billion each year needed for biodiversity protection and management.

Digital payment for ecosystem services (PES) schemes may offer a solution to this problem. PES can be built around environmental stewardship in areas such as forest conservation, wetland mitigation, water quality, and many more besides. These schemes are intended to compensate communities for positive outcomes for wildlife and ecosystems, and/or offset the costs borne by those on the front line. A prominent example of PES is carbon sequestration payments, which incentivize forest owners to keep forests intact to store the sequestrated CO_2 (e.g. the UN REDD program). Another example is wildlife conservation incentive payments, which reward landowners for maintaining biodiversity.

PES programs offer a twin promise of tackling the drivers of environmental degradation and land use change by costing the value of ecosystem services, while also generating alternative sources of income for vulnerable communities, primarily in the global south. As of 2018, there were over 550 PES programs in operation, generating USD30–50 billion in transactions. [25]

Two of the common challenges faced by PES systems are the verification of the quality of the ecosystem service, and the handling of payments to a distributed group of people. This is another area in which digital technologies may offer a solution. [26] The Blockchain Ecosystem Payments project focuses on using satellite images, combined with smart contracts on the blockchain, to pay for biodiversity corridors. The app automatically measures the wildlife corridor at regular intervals, and if the forested area has stayed within the agreed-upon limits, the blockchain smart contract will automatically trigger payments to local community members on their mobile phones. Such software could radically improve the speed, traceability, cost efficiency, and transparency of PES schemes. [27]

Another growing application is the ability to incentivize data collection by local communication via micropayments, as blockchain-enabled decentralized transfers, something that can be done at little cost even for people without bank accounts. Community-based, on-the-ground efforts can validate data collected from space, or fill key gaps, such as data from beneath the tree canopy, where satellite imagery cannot penetrate. Micropayments can help provide income for local populations, while at the same time increasing awareness and shifting the value of biodiversity for those populations, from what can be extracted to what can be observed and reported.

A few early players in the market are trying to develop exchanges for these tokenized data sets. One of these is "Proof of Impact" – an impact marketplace platform that brings together suppliers with the users of impact data. Impact sellers provide multiple data points across different data layers that are verified by the platform to generate impact tokens that prove impact.

Another of these solutions is the Earth Bank of Codes. This solution aims to put all genetic codes of the biodiversity of the Amazon rainforest on the blockchain. Pharmaceutical companies and scientists will then be offered the opportunity to buy access to the genetic information, using cryptocurrency, which is programmed to be directly paid to the communities taking care of the rainforest. It recognizes indigenous bio-IP ownership, and shifts how value is extracted from nature. [28]

However, translating our environmental goods into value is only one side of the coin. The transformational force of Valuing can also play an integral role in increasing transparency and accountability with respect to transactions occurring in the carbon market.

The role Valuing plays in transforming the carbon market

The concept of the broader carbon market involves governments and regulatory authorities seeking to cap greenhouse gas emissions related to sectors, industries and organizations. There are already several carbon markets in operation, and typically they are based on how many metric tons of CO_2 are emitted. Those that emit less than their allocation can then sell their credits to other businesses, thus incentivizing low carbon emissions across entire industrial sectors, and providing competitive advantage to the organizations that achieve this. An example is the Carbon Offsetting and Reduction Scheme for International Aviation, associated with the aviation industry. [29]

Carbon credits are an important mechanism in the carbon market, sometimes also referred to as carbon offsets. These credits essentially represent a permit system that enables organizations that cannot immediately or viably reduce their carbon emissions, to purchase credits allowing them the permission to emit greenhouse gasses (GHGs). For example, one carbon credit grants permission to emit one ton of carbon dioxide. [30]

Carbon credits can also be traded between those who have an excess of them and those who need to purchase extra credits to enable permission for them to emit GHGs beyond their imposed limits, providing a further incentive for companies to reduce emissions. [31] Carbon credits have sometimes been controversial, with some suggesting that the scheme hasn't always been successful in its aims, but at its core it is intended to produce measurable, verifiable emission reductions.

Carbon credits are, without doubt, complex instruments that are often very opaque. This can often make it difficult for markets to accurately assess their value or track and trace how they are being traded and at what price. Without key price data to value a project, a financier with a mandate to invest in sustainability-driven assets will struggle to assess the worthiness of an investment.

Viridios AI is one organization that has applied the transformational force of Valuing to develop a groundbreaking approach to solving this problem.

Through the computing power of AI, with only a few thousand historic transaction data points, they were able to train models to accurately price and value the attributes and taxonomy of a carbon project, including any combination of the UN SDGs.

With this data now available via a subscription and the publishing of CARBEX, their co-branded indices made available through their partnership with S&P Platts, market users can now see how prices change for different carbon projects and the premiums and discounts mapped to the quality and impact of the SDGs in those projects.

For example, revenues from an avoided deforestation carbon project in Cambodia have not only stopped a rainforest on the brink of destruction avoiding carbon emissions, but have also provided the funding to protect endangered species (SDG 15), and create jobs and much needed infrastructure for the local communities, ticking off SDGS 1–6 as well. [32]

By placing a value on each of these co-benefits, a project developer can now model a return on investment by knowing exactly how to translate environmental value into its economic equivalent. This then enables the deployment of more funds in building schools, providing scholarships and hiring rangers to protect endangered local animals, for example.

The benefits of this technology are also supporting the buy-side of this marketplace. For example, the corporate buyer, with publicly disclosed sustainability targets, needing to make carbon offsets to mitigate surplus emissions in other areas of their balance sheets, can know how much to pay for projects with specific additional co-benefits to help solve a particular SDG target that is underrepresented, such as biodiversity. The greater the flow of funds, the greater the incentive to produce more sustainable business and projects, exactly how capital markets should work.

Also of note is the carbon credit marketplace that Salesforce has just launched. The company has recently announced a unique carbon credit marketplace, which empowers organizations to take climate action. [33] Salesforce proclaims that the new platform will connect buyers with entrepreneurs that have ecological credentials and is intended to usher in a new era of transparency in the market, in the process aggregating third party ratings and pricing for projects that reduce carbon emissions.

Technologies such as those developed by Viridios AI and Salesforce that harness the transformational force of "Valuing" act as a key driver of other aspects of the carbon economy as well, such as the niche industry that is the voluntary carbon market. The voluntary carbon market is aimed at inducing investment in activities that reduce greenhouse gas emissions. This is achieved by enabling parties that emit carbon to offset unavoidable emissions by purchasing carbon credits. [34] This process can take place voluntarily, or as part of several schemes that operate across industrial sectors as discussed previously in this section.

Valuing can transform this industry from its current several billion US dollar annual turnover to potentially trillions of US dollars in the next decade. An industry of that size can make a significant and meaningful contribution toward achieving the 2030 UN Sustainable Development Goals.

Summary

Through these examples, I hope to have given you a glimpse of the potential for Valuing to revolutionize sustainability and overcome some of the major sustainability barriers in our economic operating system. If the force of Valuing starts to scale for sustainability, we will see massive transformational outcomes, including the following:

First, a proliferation of data and digital infrastructure that underpins the trade of goods and services will be used to drive forward a circular economy. Materials will be tracked across their lifecycle, and successfully recovered, repurposed, recycled or re-used. Digital infrastructures will also support the transition from turning products into services.

Second, real-time environmental and carbon performance of supply chains of individual products and services will be contained within digital product passports and be able to seamlessly compare it with other products.

Third, there will be a massive diversification of markets, products and services to match value preferences between buyers and sellers. The intention-action gap will be closed, as companies become more capable of communicating their sustainable value proposition to consumers, enabling consumers to become more aware of how much impact their decisions can make, and how easy it will become to make more sustainable decisions.

Finally, finding new business models such as the carbon market to align organizations, products and services with the values and goals embodied by SDGs opens an economic prize of at least USD12 trillion and potentially up to USD50 trillion by 2030, for the private sector, as recently estimated by the World Economic Forum. Indeed, harnessing the transformational force of Valuing is not only urgently needed for sustainability's sake but, put plainly, also makes perfect economic sense.

And yet to achieve these transformative outcomes, many enablers are still needed.

First, as so often in the tech world, data and standards are key enablers, but there are still areas of uncertainty. What exactly does it mean to be sustainable, and which metrics should the industry use? Who sets the data standards? What technology should be involved? National governments, and leading industry and academic groups, must collaborate on common approaches to standards, including on data quality, interoperability, carbon accounting, lifecycle assessment, disclosure frameworks and security. Greater access to protocols between businesses, governments, and consumers will also be essential. Using a common language and structure, including common semantics, ontologies, and taxonomies is a prerequisite to obtaining the level of efficiency and data-sharing needed to enable a circular economy.

The second enabler is government support and policy. For example, the UK government's first Green Finance Strategy, published in 2019, describes the accelerated growth of green finance as being vital to the country's economy. Greening the global financial system and catalyzing investment must drive innovation in financial products, in the process building skills across the financial services sector. ESG frameworks need to be enabled, with access to real-time environment and climate data being included as part of spatial financial frameworks.

The final enabler of Valuing will be the guidelines, rules and regulations from regulators. These will help guide the innovation of large and small tech companies alike and will shape how the green finance sector will evolve over the coming years. Green finance may be a rapidly growing space, but with the appropriate enablers in place, green fintech can play a critical role in building, or at least financing, a better planet. But this will only occur once our whole concept of what represents value evolves appropriately.

References

1. Schonberger, V. & Ragme, T. (2018). *Reinventing Capitalism in the Age of Big Data.* John Murray.
2. World Economic Forum. (2019). *Our Shared Digital Future Responsible Digital Transformation – Board Briefing.* Available at www.weforum.org
3. GS1. (2022). *Circular Economy & Sustainability.* Available at www.gs1.org
4. GS1. (2022). *Traceability.* Available at www.gs1.org
5. Saiidi, U. (2018). *Inside Alibaba's new kind of superstore: Robots, apps and overhead conveyor belts.* CNBC.
6. Everledger. (2022). *Diamonds.* Available at www.everledger.io
7. Everledger. (2022). *Batteries Life Cycle Management.* Available at www.everledger.io
8. Commonwealth Bank of Australia. (2022). *Australian Fixed Income Investors' Perspectives on the Green, Social, Sustainability (GSS) Market.* Available at www.commbak.com.au

9. Walker, L. (2017). *This New Carbon Currency Could Make Us More Climate Friendly*. World Economic Forum.
10. Chen, M. (2021). *Sustainable Investment Funds Near $2 Trillion in Assets*. Available at www.nasdaq.com
11. GeSI. (2022). *Mission & Vision of GeSI*. Available at www.gesi.org
12. Herden, C. et al. (2021). "Corporate digital responsibility." *Sustainability Management Forum | Nachhaltigkeits ManagementForum* 29: 13–29.
13. European Commission. (2022). *International Public Procurement*. Available at www.europarl.europa.eu/
14. United Nations Conference on Trade and Development. (2020). *Global e-Commerce Hits $25.6 trillion – Latest UNCTAD Estimates*. Available at www.unctad.org
15. GrowGlobal. (2020). *Global E-commerce: B2B E-commerce 5 Times Bigger Than B2C E-commerce*. Available at. www.growglobal.com. Accessed June 12.
16. BusinessWire. (2019). *SAP Ariba Opens World's Largest Business-to-Business Network to Connect With Swisscom's Conextrade*. Available at www.businesswire.com. Accessed October 23.
17. BusinessWire. (2019). *SAP Ariba Integrates with Givewith Platform, Enabling Buyers and Suppliers to Drive Social Impact Through B2B Transactions*. Available at www.businesswire.com. Accessed November 18.
18. PRNewsWire. (2018). *New Social Impact Technology Company "Givewith" Enables Businesses to Increase Sales and Profits While Giving Millions to Nonprofits*. Available at www.prnewswire.co.uk
19. Mastercard. (2022). *Turn Purchases into Meaningful Action: A new way to look at environmentally informed spending*. Available at www.mastercard.com
20. Walsh, M. (2022). *FinTech goes green*. Lexology.
21. Eltobgy, M. et al. (2021). *Here's Why Comparable ESG Reporting is Crucial for Investors*. World Economic Forum.
22. Chestney, N. (2021). *Global Green Bond Issuance Hit New Record High Last Year*. Reuters.
23. Haahr, M. (2019). *New SDFA report on Digitization of Green Bonds. Sustainable Digital Finance Alliance*. Available at www.sustainabledigitalfinance.org, Accessed September 28.
24. Green Digital Finance Alliance. (2020). *Fintech for Biodiversity: A global landscape*. Available at www.naturefinance.net. Accessed March 1.
25. Moros Cañón, L. et al. (2019). *Payments for Ecosystem Services in Colombia: discourses, design and motivation crowding*. Ecological Economics.
26. Neligan, P. & Quigley, T. (2020). *Next gen payments for ecosystem services*. Environmental Defence Fund. Available at www.medium.com. Accessed April 4.
27. Ibid.
28. Green Digital Finance Alliance. (2020). *Fintech for Biodiversity: A global landscape*. Available at www.naturefinance.net. Accessed March 1.
29. International Air Transport Association. (2022). *Aviation Carbon Offsetting: Guidelines for Voluntary Programs*. Available at https://iata.org
30. Kenton, W. (2022). *Carbon Credits and How They Can Offset Your Carbon Footprint*. Available at www.investopedia.com. Accessed August 19.
31. CarbonCredits.com. (2022). *The Ultimate Guide to Carbon Credits and Carbon Offsets*. Available at www.carboncredits.com

32. Milne, S. (2012). "Grounding forest carbon: Property relations and avoided defor-estation in Cambodia." *Human Ecology*, 40: 5 (October): 693–706.
33. Salesforce. (2022). *Salesforce Announces First-of-Its-Kind Carbon Credit Marketplace, Empowering Any Organization to Take Climate Action on Their Journey to Net Zero*. Available at www.salesforce.com. Accessed September 20.
34. Blaufelder, C. et al. (2021). *A Blueprint for Scaling Voluntary Carbon Markets to Meet the Climate Challenge*. McKinsey Sustainability.

7 Transformational Force #4

Embedding

Defining the force

Embedding refers to the systematic integration of sustainability data, metrics, outcomes and values into infrastructure, code, platforms, algorithms, filters, devices and other technologies.

The force of Embedding can be applied to almost all ideas, products, services, and technology applications which include physical infrastructure, digital platforms, AI algorithms, recommendation engines, IoT devices, software applications, technical infrastructure, cloud, digital product labels, supply chains, data value chains, feedback loops, procurement processes, and many more besides.

Embedding can be applied at any stage of the design, build, execution or evaluation phase of a product or service. For example, by embedding code, algorithms or technologies, that are focused on sustainability during the design phase, embedding can **optimize** the impact of sustainability solutions. Embedding can also **reinvigorate** existing technology, products or services, as well as established infrastructure, so that they are capable of the **innovation** needed to meet developing sustainability standards; in the process, ultimately extending their lifecycle. Embedding can also help **automate** or **simplify** the purchase or procurement of a sustainable product or service, according to a comparison of best performance. Embedding such performance data about a product or service, as part of its digital profile, can then enable the **interoperability** of its performance metrics across the entire supply chain. This **exchange** of information then **catalyzes** a **circular** economy model whereby provenance, supply chain dynamics and disposal or recycle pathways can be **translated** into real-world impact.

Hardcoding sustainability into the digital economy

The fact that all transactions in the economy are becoming digital means that they are all supported by digital code and algorithms. [1] This can be used as a pathway for exponential transformation if it can be harnessed in a manner where it begins to optimize and favor sustainability outcomes.

DOI: 10.4324/9781003187523-9

It could be argued that business as usual has been extremely successful at harnessing the force of Embedding for some time, given that, for the most part, digital algorithms have been embedded with a core purpose of maximizing profit, and have typically capitalized on our human weaknesses and desires to do so. In fact, Reid Hoffman, the founder of LinkedIn says that most digital business models are based on embedding the seven deadly sins into them. When these became encoded at the algorithm level, it can be no surprise that we see outcomes such as social division, the amplification of fake news, hyper-consumption, etc.

How much of a game changer would it be then, if the force of Embedding were to be used with the same voracity to integrate values such as trust, generosity, transparency, sustainability, equality and more? That is why I have identified Embedding as a transformational force – because it can play an essential role for sustainability in this way across the three levels of digital transformation. Recall:

Level 1 (Products/Services) Embedding sustainability data, metrics, values and technologies into the supply chain and design of products, services and infrastructure

Level 2 (Brakes and Accelerators) Embedding sustainability data, metrics, values and technologies into operating systems

Level 3 (Digital Ecosystems) Embedding sustainability data, metrics, values and technologies into the digital ecosystem capable of interfacing with data and platforms to drive planetary-scale outcomes

If Embedding can be applied with regularity at all three of these levels, we will finally be able to embed sustainability into all products, services, processes and so much more. Major shifts in our operating systems will result in driving and reinforcing planetary sustainability, resilience, circularity and regeneration.

Embedding Sustainability into our products and services

Embedding is changing the way that consumers perceive existing brands and forcing them to evolve in what is a rapidly developing and maturing marketplace.

One example of this is how Embedding as a force is changing the experience online customers are having when they engage with their shopping carts turning their end of sale transactions into opportunities to make impact.

Iequalchange have developed technology that thousands of retailers are choosing to embed in their checkout process that prompts their customers to donate to a cause they are passionate about. When a customer has finished shopping and chooses to check out, they are prompted to decide whether they would like to participate in making an impact and are given a choice as to what worthwhile cause they wish to support. Customers can also see what impact donations to the cause have made so far. Retailers are also able to track the progress the donations of their customers have made and promote it on their website and social platforms, leading to even more opportunity for impact.

And it isn't just the consumer market that is seeing sustainability values being embedded into point-of-sale transactions. Procurement at enterprise level is also experiencing transformation in this space thanks to organizations such as GiveWith.

The GiveWith Platform enables buyers and suppliers to drive social impact through B2B transactions. The platform identifies nonprofit programs that will deliver the maximum business value, measures and tracks project outcomes based on the level of contribution given and provides detailed reporting. This enables the translation of the social impact from the transaction into hard numbers that can then be used to communicate an organization's mission and progress in their ESG reporting, as part of their marketing strategy when communicating their sustainability value proposition to their clients and if relevant, even be accounted for on their carbon emissions balance sheet.

Platforms such as these have a flow on effect that compounds their impact on sustainability substantially because B2B supply chains are never one sided. For example, each organization is usually engaged as both a buyer and supplier in their procurement processes. In many instances a buyer will transact with the GiveWith platform as a customer of another organization but then also embed the same technology in their end of sale process for their own buyers – perpetuating a cycle of impact throughout entire supply chains.

These examples demonstrate how the force of Embedding can be used to digitize, automate, simplify and optimize sustainability values within products and services that may not have otherwise had anything to do with sustainability at all, giving everyone an opportunity to participate in outcomes that support sustainability. And by seamlessly managing the distribution, monitoring, tracking and reporting of contributions and their impact, the process is virtually frictionless. This means that not only is the opportunity massive for such solutions to amplify rapidly but no-one has an excuse not to participate either.

However, in my opinion, the most powerful impact such use cases of Embedding make for our people and planet comes from the fact that they also digitize, automate, simplify and optimize what is arguably one of the better characteristics of humanity – generosity. Imagine the impact we could make for the planet if all transactions made became a digital expression of this trait. Huh. And these are but two examples. So many more exist. Indeed, the force of Embedding is a powerful ally for sustainability that provides an exponential opportunity to change how we engage with products and services to catalyze change by enabling the digital expression of the best aspects of humanity in the process.

Embedding Sustainability into our infrastructure

Supply chains and consumer behavior are not the only areas where Embedding can be used as a formidable force for good. We can also harness it to make an impact on our built environment too. Urban infrastructure remains with the population for generations and is therefore a crucial component of shaping the built environment.

It is well known that infrastructure makes critical contributions to greenhouse gas emissions, with one particularly surprising culprit involved. The importance of buildings as a generator of CO_2 is often overlooked, yet they are accountable for almost 40 percent of all energy-related CO_2 emissions and 33 percent of greenhouse gas emissions. [2] The construction and operation of buildings was responsible for 13.1 gigatons of global energy-related CO_2 emissions in 2015. [3] It is therefore critically important to embed technology into infrastructure in order to achieve sustainability goals.

And there is huge potential for improvement in this field, as noted by the Coalition for Urban Transitions. Their report in 2019 indicated that it should be possible to cut emissions from cities by approximately 90 percent by 2050. [4]

Embedding smart building technology into infrastructure is one powerful way of achieving this. This innovation uses Internet of Things technology to disseminate information, enhance cooperation and communication, and control operations within homes and businesses as well as throughout the systems in any built environment such as a city. These technologies can serve sustainability purposes.

Smart building technology offers several useful applications that address the worrying reality that buildings consume vast amounts of energy, while also generating excessive greenhouse gas emissions.

First, smart buildings can help deliver increased energy efficiency; for example, Embedding Internet of Things technology into smart buildings ensures that numerous elements can be monitored and improved. In the UK alone, it is estimated that an increased exchange of data in relation to infrastructure could deliver an additional £7 billion in revenue annually, which could then be invested in future-proofing buildings for sustainability. [5] The data collected also helps increase sustainability throughout urban infrastructure via building automation systems. Tracking energy consumption by area, operating intelligent temperature controls, implementing humidity tracking, using sophisticated CO_2 monitoring, and vastly improving air quality have become feasible as a result of digital smart building technology.

Embedding smart grid technology into infrastructure can also play a role in ensuring that it is more flexible and sustainability focused. And there are positive trends in this area. A study led by ASI Thought Laboratory indicated that SMART water meters and SMART grids have an adoption rate of approximately 70 percent globally. [6]

Embedding building information modeling (BIM) into the planning a design phase of construction projects is also beginning to prove it has a beneficial effect on the environmental impact of buildings. BIM is a form of modeling technology that can be used by architects and designers to create a 4D digital twin of a proposed building or infrastructure project. This software can be used on anything from domestic housing to huge skyscrapers and even town planning projects. It can provide insight into everything from heating, ventilation, and air conditioning (HVAC) to all aspects of the floorplan as well as complex plumbing, electrical and mechanical grids, and more. It can also be embedded with geo-spatial technology to enable the understanding of how various buildings and infrastructure projects would interact with each other once they are built and in doing so can flag some of the unintended consequences of moving forward with the construction of such projects.

The design stage is critical to the construction of eco-friendly buildings and infrastructure projects, and therefore BIM can play a major role in sustainability going forward. Implementing BIM can also make it more likely that buildings are able to obtain sustainability certifications, such as the Leadership in Energy and Environmental Design (LEED) certificate.

Two successful examples of the implementation of BIM technology come from China. The application of BIM in the "China Beautiful Rural" design project in the Yangyou village river ecological landscape reconstruction design project has been highly acclaimed, [7] while the Chinese authorities have also used BIM technology to analyze landscape planning and design to improve the living standards of rural citizens. [8]

Of course, the major opportunity for the force of Embedding in this respect is to utilize BIM technology and embed it in larger digital twin models that give feedback on how building and infrastructure projects will impact on the natural systems and ecological landscapes of the surrounding areas as well as its citizens. This is where smart cities come into the picture.

Smart cities utilize technology like predictive AI to achieve better outcomes. By leveraging the force of Embedding through the use of Internet of Things connectivity combined with agile artificial intelligence and machine learning, smart cities can become intelligent and predictive, addressing problems before they even arise. Predictive analytics can be utilized to track energy requirements, control pollution and assist with more cohesive public transportation systems and intelligent waste management and recycling.

Above all else, smart cities harnessing the transformational force of Embedding to put sustainability values at their core, can assess their impact on the environment, with machine learning enabling authorities to make informed decisions that ensure our cities remain in line with critical ecological and sustainability goals. Perhaps then we will finally be able to see into the crystal ball and prevent outcomes such as contributing to the "Urban Heat Island Effect" (recall, we covered this in Chapter One) and other unintended consequences our cities have on the planet.

Embedding sustainability into the next generation

One area of popular culture that undoubtedly possesses untapped potential for harnessing the transformational force of Embedding, is the gaming industry. Gaming was one of the few entertainment sectors that was barely impacted by the COVID pandemic; in fact, there is evidence that the industry expanded exponentially during this period precisely because it was accessible. For example, Nintendo reported an increase in profits of 41 percent while another popular gaming platform reported a 31 percent increase. [9]

Thus, PwC's Global Entertainment and Media Outlook 2022–26 report suggests that the gaming industry will be worth USD321 billion in 2026. [10] The growth in gaming is also being fueled by platforms such as Twitch, which have attracted vast audiences. 22.8 billion hours of mostly gaming content was consumed on Twitch in 2021, and the concept of broadcasting games continues to grow in popularity. [11] Indeed, using the force of Embedding to turn the humble computer monitor, SMART device or television screen into an unlimited source of entertainment has been a gold mine of epic proportions.

This has raised the eyebrows of the many actors focused on sustainability solutions. They have recognized that Embedding games with positive sustainability messaging can provide a fertile platform to incentivize the 3.24 billion 'gamers' that exist globally to become more aware of how their choices impact the planet. [12] Gaming is, unsurprisingly, popular with our younger generations, with 79 percent of people under 22 years of age being classified as 'gamers' in the UK and in the US it is estimated that 38 percent of all gamers are 18–34 years of age. [13] It is therefore not only an important audience for the future but also a captive one.

Of course, the digital big brother of nudging, as discussed in depth in Chapter Five, takes a leaf out of the gaming book by using a technique called gamification. This refers to the use of game mechanics and game design techniques in a non-gaming context. This includes embedding points, badges and other rewards, unlocking levels, and leaderboards. It's a powerful tool to engage employees, customers, and the public to change behaviors, develop skills and drive innovation.

Examples of this interactive approach are proliferating across a variety of sectors.

For example, Angry Birds 2 held an in-game event to raise awareness for climate change and reforestation. For a limited time, players were able to collect a new Forester hat set and adorn their flock of characters in some rugged new clothing. In addition, a community event enabled players to band together and show their support, by popping as many pigs as possible. The events directed players to the United Nations Trillion Tree Campaign, an initiative to plant (you guessed it) one trillion trees by the end of the decade.

Within the Playing for the Planet Alliance, 29 video gaming companies, with a combined reach of well over 1.2 billion monthly active users, are working to embed authentic green activations in games. Together, they are exploring the power of gameplay to raise awareness on environmental issues and individual behaviors in subtle ways that are an organic part of the core game narrative. Indeed, as observed by Sam Barratt, one of the founders of the initiative, "*the facets of leveling up, going faster, scoring higher and taking on the impossible, are all critical ingredients we need to gather from gameplay and direct into the greatest multiplayer challenge of our time: climate change*". [14]

This work has also extended to the Green Games Jam, where gamers develop and embed environmentally themed games on major platforms. In 2021, 377 submissions were made by gamers on the Sony Dreams platform.

Some fitness apps are also getting into the business of Embedding gamification into wearable devices and the apps they speak to by linking fitness with environmental outcomes. Every year, Adidas and Runtastic host Run For The Oceans – a virtual challenge where participants can run, walk, or wheelchair to help support Parley's Global Cleanup Network. The Parley's Global Cleanup operation works to help end marine plastic pollution by intercepting debris from beaches and islands. For every kilometer run between May 28–June 8 via the Adidas Running app (Strava/Joyrun), Adidas and Parley will clean up the equivalent weight of 10 plastic bottles, up to a maximum of 500,000lbs of marine plastic waste from beaches, remote islands, and coast lines. To-date, 5 million people have taken part in the competition, running a total of 56 million kilometers. [15]

Another example is UbiGreen. A mobile phone application which semi-automatically senses your means of transportation, and provides corresponding information on this behavior, indicating the CO2 emissions that are caused by each mode of transport you could choose to take. Small rewards are given to those who make 'green' choices. Feedback is provided over two different interfaces: One shows a tree, and the other a polar bear on a small iceberg. When green means of transportation are chosen, the tree grows more leaves, and in the final stage polar bears receive more apples. Correspondingly, the iceberg grows bigger and harbors more animals. [16]

WeSpire also created a platform where users earn points for completing certain sustainability actions, such as recycling or using more environmentally friendly products. Points are shown on a leaderboard, and achievements can even be posted on Facebook, turning the persuasive power of social media into a tool for positive change. The platform has attracted a great deal of buzz, with companies such as MGM, Sony and McDonald's all using it to launch their own sustainability challenges, and six million positive actions taking place to date.

A third and final gamification technique to hack our own sustainability is through subtly embedding sustainable products, services, behaviors and memes in our various forms of digital entertainment and social media. [17] This involves trying to normalize sustainability choices in the behaviors and actions of characters, plots and outcomes in a way that feels authentic, lived-in, and believable (and without directly preaching).

Indeed, the gaming industry represents a worldwide platform of billions of users that can become an opportunity to raise awareness and make sustainable practices more mainstream. This can happen through the content on screen, as well as through their own practices behind the screen. This kind of green product placement or 'positive placement' can help to modify an individual's beliefs, attitude and behavior with repeated exposure.

While an important element of Embedding is addressing how humans engage with sustainable messaging, it is not limited to this. Another important aspect of the Embedding concept is changing 'what' we use, to ensure that products engage with our sustainability goals in mind.

Man and the machine

As technology continues to evolve, and innovations such as AI and machine learning harness the force of Embedding to infiltrate more and more of society in general, the rise of robotics as a sustainability solution is inevitable.

At the time of writing, Tesla has recently unveiled its Optimus robot, asserting that the human-like machine will be "amazing" within 5–10 years. [18] Optimus has its supporters and critics, as with many Tesla projects, but the commitment of the company to robotics is indicative of a trend that will become more important in the coming years. And so we have to ask how such machines being embedded with technology that is likely smarter than its human operators will engage with our planet on our behalf? Embedding sustainability code and algorithms into our robots could indeed be critical for achieving sustainability goals.

Most technology experts now believe that the integration of robots into our daily lives is inevitable; all that remains to be decided is the timeframe. This will unquestionably represent a paradigm shift in human society, a line in the sand that has been depicted in science-fiction for many decades. But it can also be a massive positive for the environment. Indeed, robots are already being used for tasks that have traditionally been human, in a way that has ecological benefits.

The field of miniature robotics, or micro-bots, has already provided sustainable functionality. In a recent study, researchers from Keio University embedded sustainability software solutions into micro-bots. They created color-changing microrobots to freely explore and gather information on the environment. The tiny robots, less than 1mm in diameter, sit on the surface of lakes and rivers, detecting changes in the properties of the water. It is hoped that the project will eventually be able to detect heavy metals, explosives and environmental toxins. [19]

Micro-bots are also being embedded with technology to mimic the role bees play in the pollination process of plants. When we use technology in such ways, to mimic the behaviors of things in our natural environment, it is called biomimicry. Stickbug is a project from West Virginia University that is aiming to develop micro-bots that can pollinate plants in an indoor environment. This use of computer vision algorithms to map out static environments can help train micro-bots to identify flowers that need pollinating and stands to revolutionize the vertical, indoor and greenhouse farming industries. The upside of such projects for sustainability could be huge.

Drones that are embedded with geo-spatial, cadastral imaging, camera visioning and AI technologies that enable them to plant trees are already capable of seeding 100,000 trees in a single day – one of the most productive ways of reducing CO_2 in the atmosphere and a significant opportunity to make massive inroads into reforestation, regeneration and biodiversity projects. Biocarbon Engineering, a startup based in Oxford, has been operating its drones in Australia, planting trees and grasses in an array of abandoned mines. [20]

The same technology is also being used to embed programs in drones to make them effective at environmentally friendly weed and pest control, the micro-dosing of plant nutrients and precision application of fertilizer as well as saving lives and biodiversity by assisting the fighting of wildfires.

Robotics is proving helpful in other fields as well. Using the force of Embedding to program robots to sort waste materials more quickly than humans ever could, assists with recycling and waste management.

Robots are also already assisting on the ground as first responders alongside emergency personnel with the fight against wildfires in the United States, having been successfully armed with fire extinguishers and water propelling agents.

Harnessing the force of Embedding to program robots with the code and algorithms to perform menial and/or dangerous tasks in such fields as renewable energy and sustainable agriculture, means performing these tasks faster while keeping humans safer and freeing them up to work on more skilled and less onerous tasks. [21]

Embedding Sustainability into regulation and policy making

While the benefits associated with the force of Embedding are obvious, it, like every other use case associated with the five forces, poses moral, ethical and practical considerations. In accordance with this, it is essential to implement regulation and policy related to sustainability and at a global level. Central to this process will be harnessing the force of Embedding and its related technologies to embed sustainability data, values, metrics, algorithms and technology into policy creation and regulation as well as the legal frameworks that govern our organizations and regulatory bodies. This process can have value at the local and national level too but creating a sustained global effort will require a unified approach.

Climate Policy Radar is a UK-based not-for-profit organization harnessing the force of Embedding to support evidence-based decision-making in relation to climate policy.

This philanthropy-backed climate data startup began developing a platform in 2021 and is utilizing artificial intelligence and data science to build the world's largest, most comprehensive open knowledge base for climate laws, policies and litigation, effectively creating a one-stop shop for structured, big data about climate change strategies in countries, states and cities globally.

By automating the aggregation, organization and analysis of this type of climate data they are turning unstructured lengthy documents into meaningful, accessible knowledge for all. For Climate Policy Radar, the name of the game is to create a world first global climate policy landscape. As their founder and CEO, Michal Nachmany explains, "*This enables stakeholders like governments, researchers, organizations and civil society the ability to efficiently explore policy options and identify best practices. This supports decision-makers to forge evidence-based pathways to reduced emissions and enhanced resilience that work in their local contexts*".

With greater access to useful information such stakeholders can then collaborate with intergovernmental organizations, regulatory bodies and the legal community to track the impact of policies and challenge leaders over their actions, or lack thereof.

The possibilities for other applications of this platform are massive as the data and insights can also then be embedded into strategic policy metrics to be built into risk models, guiding, among other things, climate finance, investment, lending decisions, resource management and the planning and impacts of large-scale infrastructure and more.

While this is just one example of how organizations are utilizing the force of Embedding to ensure sustainability values and metrics can be embedded into policy creation, regulatory frameworks and more, the mind boggles as to what would be possible if we chose to do this across every aspect of policy and regulation that touches the SDGs.

Summary

Through these examples, I hope to have given you a glimpse of the potential for Embedding to revolutionize sustainability and overcome some of the major sustainability barriers in our operating systems. If the force of Embedding starts to scale for sustainability, we will see massive transformational outcomes, including the following:

First, we will see the Embedding of sustainability data, values, metrics and technology enable brands and organizations to weave support for sustainability outcomes into their business models and product offerings. This will create opportunities for brands to direct their loyal customers to additional opportunities to act for people and the planet.

Second, we will see a wave of sustainability focused software applications and algorithms embedded into apps and games, effectively turning smartphones, tablets, gaming stations, wearables, sensors and a wide range of intelligent devices into tools to be used by the next generation so they can act as agents of change for the planet.

Third, by embedding sustainability into the codes and algorithms that dictate how we use such innovations as robotics, we will craft a world where sustainability can be automated, simplified and optimized beyond our wildest imagination.

Last, digital governance mechanisms enabling change across systems, organizations, industries and nations to occur at the speed and scale, and with the transparency and accountability needed can finally be realized.

And yet to achieve these transformative outcomes, many enablers are still needed.

Universal interoperability standards enable the operational processes that underpin the sharing of information between various systems to be optimized, as well as implementing a set of common expectations associated with them. This is critical for sustainability, as the concept presents the opportunity to disseminate information and data across multiple territories; vital if effective global frameworks for the consumption of goods and services are to be created.

Creating standards around sustainability optimization will also be beneficial, particularly as these two concepts have become inseparable in recent years. The primary goal of optimization is to improve social, economic and environmental sustainability, which can be achieved by applying optimization algorithms and data analysis techniques to collated information. We cannot embed sustainability into the codes and algorithms of the machines we use or the products and services we buy without it. Enshrining regulatory standards will again help encourage a more unified and cohesive response to the climate crisis.

And sustainable design standards encourage negative impacts on the environment to be diminished during the design phase of products, buildings and infrastructure. This can also be encouraged with incentives, such as taxes and charges, eco-design standards, extended producer responsibility, and environmental product labels.

Encouraging the principle of Embedding as a force for good across a myriad of aspects of our society will not only have positive outcomes for the environment, it will also play a role in evolving healthy social discourse. If we can harness the transformational power of the force of Embedding, eventually, sustainability will become ingrained in our operating systems, which can be nothing but positive for our collective future.

References

1. Khan, A., et al. (2019). *Linking Sustainability-Oriented Marketing to Social Media and Web Atmospheric Cues*. MDPI.
2. Tricoire, J. (2021). *Why Buildings Are the Foundation of an Energy-efficient Future*. World Economic Forum.

3. United Nations Environment Programme. (2020). *Building Sector Emissions Hit Record High, But Low-carbon Pandemic Recovery Can Help Transform Sector – UN Report*. Available at www.unep.org
4. Coalition for Urban Transitions. (2019). *Climate Emergency, Urban Opportunity: Executive Summary*. Available at www.urbantransitions.global
5. World Business Council for Sustainable Development. (2021). *Digitalization of the Built Environment: Towards a More Sustainable Construction Sector*.
6. Deloitte. (2022). *Smart and Sustainable Buildings and Infrastructure*. Available at www.wbcsd.org. Accessed February 4.
7. Wei, X. et al. (2019). The application of BIM in the "China Beautiful Rural" Design Project – Yangyou Village River Ecological Landscape Reconstruction Design Project, in *Advances in Human Factors in Architecture, Sustainable Urban Planning and Infrastructure*: Proceedings of the AHFE 2019 International Conference on Human Factors in Architecture, Sustainable Urban Planning and Infrastructure, July 24–28, Washington D.C., pp. 135–145.
8. Bai, Y. (2021). *Research on Rural Landscape Planning and Design based on BIM*. August. Journal of Physics Conference Series. Available at https://analyticsinsight.com. Accessed October 31.
9. Trends, M. (2020). Gaming Boom in COVID-19 times: Analysis & Insights. Analytics Insight. Available at https://analyticsinsight.com. Accessed October 31.
10. Read, S. (2022). *Gaming is Booming and Is Expected to Keep Growing*. This chart tells you all you need to know. World Economic Forum. Available at https://weforum.org. Accessed July 28.
11. Iqbal, M. (2022). *Twitch Revenue and Usage Statistics (2022)*. BusinessofApps. Available at https://businessofapps.com. Accessed September 6.
12. Bojan, J. (2022). *Gamer Demographics: Facts and Stats About the Most Popular Hobby in the World. August 2022*. DataProt. Available at https://dataprot.net. Accessed August 2.
13. Ibid.
14. United Nations Environment Programme. (2020). *Playing for the Planet Annual Impact Report*. Available at www.unep.org
15. Adidas. (2022). *Run the Oceans*. Available at www.adidas.com.au
16. Kolpondinos-Huber, M. (2015). Gamification and Sustainable Consumption: Overcoming the Limitations of Persuasive Technologies, in *ICT Innovations for Sustainability*, pp. 367–385. Springer International Publishing.
17. Fortuna, C. (2019). *Green Product Placement Is Making Sustainability Cool & Normal*. CleanTechnica. Available at https://cleantechnica.com. Accessed September 18.
18. Chaturvedi, A. (2022). *Know Why Tesla's Robot Optimus is a Game Changer*. TechGig. Available at https://TechGig.com. Accessed October 13.
19. Yoshida, K. and Onoe, H. (2022), "Marangoni-Propulsion Micro-Robots Integrated with a Wireless Photonic Colloidal Crystal Hydrogel Sensor for Exploring the Aquatic Environment." *Advanced Intelligent Systems* 4, 5.
20. Arpas-UK. (2022). *Tree Planting Drone Can Plant 100,00 Trees in one day*. Available at https://arpas.uk.
21. Allerin. (2019). *How Green Robots Are Helping with Environmental Sustainability*. Available at https://allerin.com. Accessed December 31.

8 Transformational Force #5

Adapting

Defining the force

Adapting is the action of a system, or components (such as a product, service, policy or infrastructure) within a system, that are required in order to function as needed in a dynamic environment.

Adapting as a transformational force does not diminish the effectiveness of other components within a system, nor the connective tissue between systems that are functioning to benefit, support, or sustain the whole, but rather seeks out and catalyzes the transformation of these opportunities to enhance this capability. It achieves this while also simultaneously making redundant those parts that are broken and/or reducing the effectiveness of the whole through natural selection – that seeks to actively pursue a constant equilibrium with its surroundings. The result? A more efficient, dynamic, flexible, resilient product, service, technology, component or system emerges in each and every case where it is applied. Simply put, Adapting is the transformational force that has emerged as a response to the fact that digitalized technologies have changed everything. Today's technologies are successfully interfacing with, and influencing, all the other operating systems. They are also becoming increasingly dynamic; for example, self-learning systems that attempt to actively influence the behavior of users, while self-adapting their environment accordingly.

As such, Adapting as a transformational forces catalyzes two main events to occur within systems and components of systems:

1. Self-organization
2. Emergence

Self-organization is the process via which a system changes its structure on a basis of dynamic feedback, without any external control, in order to adapt and respond to changes within its environment. Emergence refers to the existence or formation of characteristics and behaviors that emerge from the interaction of different systems and their components. These new

DOI: 10.4324/9781003187523-10

emergent features and properties can rarely be predicted based on the individual systems. As the expression goes – the whole is greater than the sum of the parts.

For example, products and services are an emergent property of individual businesses competing and cooperating for profit. The computer desktop and interface are an emergent property of different computer subsystems (e.g. motherboard, memory, CPU, video card), and so on.

Our social, economic and political systems are only made possible because of the total of smaller, individual systems working in synergy. The key challenge for a sustainable civilization is that emergent systems and their individual components are required to self-organize, and therefore adapt, to unforeseen conditions on a regular basis. We must ensure our social, economic and political systems are resilient to different shocks and stresses, while also possessing the ability to adapt to new opportunities. The transformational force of Adapting is already playing an essential role in our sustainability toolkit because it can be applied in immeasurable ways to each of the three levels of digital transformation. Let's recap them here:

Level 1 (Products/Services) The emergence of entirely new products and services as a result of business models, markets and consumer behavior undergoing the process of self-organization (bottom up).

Level 2 (Brakes and Accelerators) The emergence of decentralized governance, collective intelligence and collective action as a result of organizations, governments and civil society undergoing the process of self-organization (top-down).

Level 3 (Digital Ecosystems) The emergence of a dynamic, agile, responsive digital ecosystem as a result of business models, technologies, stakeholders and systems self-organizing to work toward common sustainability goals (Dynamic interaction).

The role self-organization and emergence plays for sustainability

Currently, our 'business as usual' civilization is built on legacy-based, rigid systems that do not have adequate flexibility, nor enough resilience to respond dynamically.

In relation to technology, legacy systems refer to any outdated computing software or hardware that remains in use that causes integration or interaction problems with newer technology. Given the pace of modern life and the speed of technological evolution, you might think that legacy systems are uncommon, but this isn't the case at all.

Often, certain sectors or activities become highly reliant on dated technology, and then when upgrades are required it can be difficult to extricate processes from their reliance on these legacy systems. Barriers to this can include skills gaps, high migration costs, and even sheer fear! A classic example of this is Adobe Flash Player, which was considered cumbersome and insecure, and yet remained central to the operation of the internet! Other legacy systems such as SAP, Net and Oracle are widely recognized to have major drawbacks, and yet remain widely utilized by many organizations and companies.

When the cost of outdated legacy systems is mere inconvenience then the issue is considered to be just an annoyance, but nothing that should be considered unduly serious. However, when there is a paradigm shifting issue that needs to be addressed, and a lack of technological development or evolution is causing bottlenecks in the response to this, the situation becomes rather more pressing.

The same analogy could be applied across all our operating systems. That is why, in Chapter Two, when we unpacked the challenges in these systems, we referred to them as the bugs and malfunctioning code. The bugs and malfunctioning code of our operating systems have been largely treated like an annoyance when put into the context of our business-as-usual existence. However, as we continue to ignore these bugs and malfunctioning code, the risks and unintended consequences for our civilization have become so dire that we are now faced with an urgent need to upgrade to newer operating systems that enable adequate flexibility, resilience and dynamic response so that our species has a fighting chance to survive.

We are literally one, or maybe two, major environmental and/or socio-economic events away from complete structural breakdown of our systems. Theoretical physicists from the Alan Turing Institute claim in a recent peer-reviewed paper published in *Scientific Reports* that there is a 90 percent chance civilization will collapse within decades if we continue our present trajectory. [1] As humanity's surroundings become more and more complex and chaotic, the urgent need to adopt the transformational benefits of this force have grown with them proportionately.

With so much of our lives being turned upside-down, and so few helpful answers coming from top-down for how we handle the present crises we see playing out across the world, there is a strong and growing contingent that have begun to believe that our civilization just might be ready for the force of Adapting to be unleashed. Can we begin moving, interacting, sharing, compiling, collaborating, contributing and dynamically responding as a whole-earth and a whole system?

I'd like to think we can. There are already a significant number of examples of how the transformational force of Adapting is resulting in exponential change. Let's explore them together now.

Adapting consumerism to shift away from product ownership toward an ecosystem of services

There is no better example of the transformational force of Adapting causing paradigm shifts across all levels of digital transformation for sustainability simultaneously, than that of the rise of the sharing economy. Conceived from an idea that people don't want or need to own everything they use. Consumers can now rent, barter, swap or borrow it from someone else who owns it or if they own it, be connected to someone who needs it – through platform technology and apps.

This phenomenon has turned multiple industries completely on their heads. The diffusion of profits and market share that used to be solely owned by large corporations are now to be shared with the everyday citizen. While that in and of itself creates certain benefits for our economy and planet, there are many other aspects of the sharing economy that make it an even bigger win for sustainability, the emergence of the car share and e-hailing industries being the first we will explore.

These relatively new approaches to transport and commuting remove the need for households to own cars and can thus have a significant impact on reducing greenhouse gas emissions. The car sharing market surpassed USD2 billion in 2020 and is expected to expand at an annual growth rate of 20 percent between 2022 and 2027. [2] Car sharing is increasingly seen as an affordable and convenient solution to commuting, while its ecological credentials cannot be doubted. Evidence indicates that car sharing reduces car ownership, travel costs, congestion and environmental pollution. One prominent example of a company using the force of Adapting in this way is Car Next Door.

Lynk & Co is another popular car sharing company which has used Adapting to enable car owners to use their mobile phones to set their own prices, schedules, pick-up and drop-off locations, making barriers to participate in the car sharing industry even lower. More people utilizing their mobile phones embedded with this kind of technology means fewer and fewer cars on the road.

Indeed, the impact of the vehicle sharing industry cannot be overestimated. A Transportation Research Board and National Academy of Sciences study found that each shared car takes approximately 15 private cars off the road. [3] This has resulted in a forecasted fall in the global sale of automobiles, with the figure for total sales falling to just under 70 million units in 2021 – down from a peak of nearly 80 million units in 2017. [4] With China being the largest automobile market in the world, both in terms of sales and production, it is particularly notable that Chinese car sales dipped for the first time in 20 years in 2018. [5] There have been fluctuations since then, but there are signs that our cultural obsession with owning motor vehicles is beginning to wane.

This reduction in the number of road vehicles is extremely important, as transport remains one of the major causes of carbon emissions. In a report published in 2021, the International Energy Agency discovered that transport was responsible for 37 percent of CO_2 emissions from end-use sectors. [6] It is therefore notable that each car share member reduces their personal CO_2 emissions by between 1,000 and 1,600 pounds every year. [7] In some cases, simply participating in car sharing can reduce annual mobility emissions by as much as 18 percent. [8]

Studies have also found that car sharing has an impact on so-called car shedding, in which participants sell vehicles that they own after utilizing a car sharing service. One study in San Francisco found that car sharing members in urban areas own significantly fewer vehicles than those who have never participated in a car sharing program. [9]

Another innovative use of the force of Adapting within this same space is e-hailing. Instead of consumers owning their own car, or, in the case of car sharing, driving themselves in someone else's car, they hail rides through an app that is embedded with geo-spatial technology to match people who need a ride with a driver nearby who can take them there. The number of e-hailing trips tripled over the last four years to the point there are now more than 14 million e-hailing trips being booked every single day, which accounts for 90 percent of consumer spending in shared mobility globally. This promotes shared mobility as a smarter way to navigate modern urban areas and, like the car sharing applications of Adapting, not only reduces the number of carbon emissions from vehicular travel but also reduces the demand for raw materials extraction, products being manufactured and therefore the amount of product in circulation; a win-win-win for the planet.

Uber is a well-known success story within this space. It has also recently taken the concept of e-hailing a step further for the planet with its UberPool service, in which riders heading in the same direction can opt to share the same car and split the costs of their trip fee in doing so. Whilst this feature was likely built into the Uber app to incentivize their customers to use Uber more often, Uber unwittingly built a positive feedback loop into their technology that allowed their users to self-organize to become more sustainable in their behavior. Nice one Uber. However, while the car sharing and e-hailing industries are both a particularly important manifestation of Adapting where sustainability is concerned, this merely scratches the surface. The force of Adapting is transforming the consumer's relationship to product ownership across so many other industries and verticals.

Innovative companies now realize that even when they are selling a product, what they are really doing is selling the functionality that product ultimately provides. It is therefore increasingly possible to use the force of Adapting to transform legacy business models that focused on the production of goods for never-ending consumption into service-based ones that turn consumers into agents of change for the planet. The sharing economy is the tangible manifestation of this.

In Shanghai bicycle sharing platforms reduced emissions by more than 25,000 tons in 2016 alone. [10]

Spotify and SoundCloud negate the need to own physical CD's and records, Audible and Kindle negate the need to own physical copies of books, Netflix and Youtube negate the need to own tapes and DVDs. Docusign negates the need for you to print contracts and other important documents on paper, GetMyBoat is the Car Next Door equivalent for boats and other watercraft, Babyquip is where you can hire your baby furniture from and Swimply will even let you hire someone else's swimming pool for an hour or a day – the use cases are simply endless.

The win for the planet is amplified because all these use cases have the ability to reach millions if not billions of people globally, that's a significant number of raw materials not being extracted, products not going to waste and carbon not being emitted.

Decentralized governance and the acceleration of collective action

It is imperative that our governance systems evolve in the same ways, and at the same rate and scale that technology catalyzes the evolution of industries, sectors and our economy at large.

This is because they (governance systems) are the glue that has enabled the connectivity between all our ways of being and doing in this world. While their mechanisms are far from perfect, governance systems support the flow and regulation of behavior, data, information, money, decision-making and transactions.

However, one of the quandaries of addressing the climate crisis is that the economic and environmental realities that propel our society are often not aligned by common values. We have created an economic system that demands continual growth, even while striving for this growth has incurred a significant degree of ecological damage.

The mechanisms that underpin the economic system are intrinsically directed towards the generation of perpetual growth, partly to address the desires of shareholders in an economic system that is, to a great extent, driven by their interests, and partly to service the mountain of private and state debt that exists today.

Unfortunately, this tends to run contrary to environmental goals, which require humans at the systemic level to make decisions that are not driven by unending growth, and therefore, almost inevitably, increased consumption.

Perhaps one of the most important outcomes for governance mechanisms of the future is to help humans achieve an accelerated way of finding consensus, bringing together divergent perspectives into a common vision and building empathy for other points of view. Sounds easy – but this challenge has been at the root of political contests since humans adopted democratic ideals and related institutions.

However, there are attempts to harness Adapting as a force to help humans achieve consensus and collective action that could be applied to our sustainability objectives. For example, pol.is is a digital platform designed to help groups express their interests and find consensus. It was used as the backend for a policy making process called vTaiwan – a platform that invites citizens into an online space for debate that is then heeded by politicians during the democratic process. This platform was designed as a consensus-generating mechanism, rather than the typical winner takes all polarization that often comes from policymaking, and has inspired younger voters and citizens in particular. [11] This platform has the potential to increase participation in achieving sustainability goals, and connect public concerns with politicians directly, in the process ensuring that political figures can garner a better impression of what people consider to be pressing environmental concerns.

The transformational force of Adapting can play a major role in a future that evolves toward distributed government, as everyday citizens increasingly address existing social problems, via such accessible and affordable technology as mobile and cloud. Over time, the arguments for a cloistered and centralized form of governance will diminish, and societal decision-making will, consequently, become both more participatory and inclusive. And there are already examples of some of these phenomena emerging. For example, Boston citizens play an active role in the running of their city, submitting photographs of infrastructure issues so that they can be addressed more quickly and appropriately. This concept of everyday citizens actively holding governments accountable will become an increasingly popular theme in the years to come.

The force of Adapting is also already playing a role in unifying ecosystems and enabling agents of change within governance systems. A notable example of this phenomenon is Icebreaker One, a web of net-zero data that has been described by the United Nations Environment Program as *"one of the top 20 agents of change in the path towards a digital ecosystem for sustainability"*.

Icebreaker One connects financial, industry and environmental data to help inform decisions across society that aim to produce net-zero outcomes. As part of its mission, the company has created the IB1 Trust Framework, which codifies processes and policies that govern data sharing, in the process creating an open and trusted network of accessible net-zero data. Among its goals, Icebreaker One is working toward *"a clear, trusted, impactful and detailed governance framework that enables government, businesses, consumers and third-party developers to understand the potential of open access to data"*.

Icebreaker One has partnered with Open Data Institute, the UK Government Department of Digital, Culture, Media and Sports, the Gates Foundation, CABi, Scottish and Southern Electricity Networks and Dgen. The company is also an active participant in the Future of Sustainable Data Alliance (FoSDA), formed in Davos in January 2020, headed by Refinitiv and the World Economic Forum, with IIF, ASIFMA, Tsinghua University, OMFIF, GFMA, Climate Bonds Initiative, FinTech4good, Everledger, Oxford University, the Spatial Finance Initiative, Satellite Applications Catapult, Finance for Biodiversity and GoImpact all joining as founding members. The goal of FoSDA is to *"identify and accelerate the reliable, actionable ESG data and related technology that is needed for improved investor decision making on the global journey to sustainable development"*.

The force of Adapting is also transforming the way in which projects involving multiple stakeholders can engage and distribute decision-making fairly and equitably.

Decentralized Autonomous Organizations, better known as DAOs are currently the most sophisticated digital expression of the force of Adapting for this purpose. The emergent properties of enabling people to self-organize through this mechanism are the equitable and fair distribution of authority, benefits, responsibility, accountability and decision-making across a group of people.

Members create the bylaws associated with this technology in an open-source manner. These bylaws are coded into a blockchain, and function without a single source of governing authority. That is why DAOs are considered to be decentralized. They are also capable of performing many functions autonomously which is just a fancy word for being able to do certain actions automatically, without outside input. Members all get access to the same information, which is verified and authenticated on the blockchain, and all get equal voting rights without the impact of outside influences being able to sway the vote. Depending on the purpose for which the DAO was founded, it can also automate functions to ensure accountability of members, distribution of profits and allocation of responsibility are all distributed fairly, equally and autonomously too.

This makes it a particularly powerful mechanism for the governance of digital public goods, open-source technologies, public spaces and forums too. It can also create positive feedback loops for social good investing among many other use cases.

ClimateDAO is an example of a social impact cryptocurrency that focuses on making a difference for climate change. Their 'activist' investment approach is considered completely unique. As members contribute to the DAO the members vote to distribute the money to purchase shares in publicly listed companies. As their share volume grows, they use their voting rights to initiate shareholder proposals and advocacy for climate change in ways that would have the listed company obligated to follow through.

Non-Fungible Tokens (NFTs), Decentralized Ledger Technology (DLTs) and SMART contracts are all other important examples of how cryptocurrency and blockchain solutions are harnessing the force of Adapting. All of these have many applications capable of making exponential change for our planet.

However, humans are no longer the only major form of sentience making decisions around self-organization, and, it could be argued, nor should they be.

Self-sustaining/perpetuating ecosystems

The big hope for the sustainability agenda is that we can begin to engineer self-organizing systems that help move humans toward more sustainable outcomes.

For the force of Adapting to have full impact, it depends on feedback loops that help channel information on the effectiveness of the system in achieving its goals. Therefore, self-organization is a function of the quality and breadth of data and analysis that is fed into those feedback loops. The major difference with self-organization in the digital world, compared to its analog cousin, is the ability to receive real-time data and analytics on multiple scales – from the local to global level.

AI systems themselves can look at real-time data and make autonomous decisions to self-organize different human systems. This is already happening in terms of traffic flow within cities, emergency response routing, energy management and water management.

Never have we had access to tools such as AI and big data, which, as discussed in our chapter on Sense-Making, allow us to extract insights, visualize trends, and predict future outcomes. Technologies that then utilize the transformational force of Adapting take the insights from these Sense-Making technologies and automatically apply them to a system, product, technology or service to adapt how the components of such relate to each other or change how they function entirely.

While these capabilities are a long way from being fully realized, the future versions of such systems would apply the force of Adapting to be able to do things such as identifying the gaps in our sustainability solutions and informing the design and creation of the components needed to fill those gaps to optimize outcomes for the entire ecosystem. Now that *is* exciting.

Last but most certainly, not least

As important as all the above applications of the force of Adapting unquestionably are, nothing in recent history within technology circles has been touted as the Next Big Thing (NBT) quite like that of Web3's most favored offspring; the Metaverse.

For those who aren't aware of this highly mooted technology, the Metaverse is considered to be a transformative system or systems that have harnessed the force of Adapting to completely evolve the way that we interact with the internet, but it could also have huge impacts on how the future of our civilization would engage with their 3D existence.

The Metaverse is inextricably linked to virtual reality (VR) and augmented reality (AR), as the model for the system predicts that it will be increasingly possible for internet browsing and communication to take on a much more three-dimensional aspect. Metaverse users would utilize VR and AR headsets to connect with what would become a hugely immersive world, in which experiences well beyond the relatively crude interfaces of the existing internet are possible. Ultimately, the Metaverse can be seen as a convergence of these digital, augmented and virtual worlds.

If ever there was a technology that was poised to revolutionize the governance, agility, self-organization and self-perpetuation of the sum total of all our systems through harnessing the full transformational force of Adapting; in my mind's eye, the Metaverse is it.

To date, no other technology had the ability to adapt every aspect of how each component of our products, services and daily lives interact, nor can we comprehend fully what additional components and properties will emerge from all of these newly created methods of engagement.

When its full potential is realized, it will be capable of ingesting all forms of technology and facilitating the self-organization of them, enabling the emergence of entirely new experiences, transforming how we transact with each other and how we engage with our lives and the world at large; on an individual, group, organization and government level. The Metaverse evolving into a digital twin of our civilization as we know it, is a likely outcome, thus, enabling, in my opinion, the largest social, economic and governance ecosystem experiment in the history of all mankind.

It should be emphasized, of course, that this is all entirely hypothetical! The Metaverse certainly isn't functional, much of the technology required to deliver it has yet to be developed, and some interested parties are skeptical that the Metaverse will ever exist in its transformative state, or at least within the foreseeable future. Conversely, other tech experts are enthusiastic about the potential of this technology, with some deeming it inevitable.

What has given the Metaverse its initial momentum is the huge investment made by tech giants like Facebook. How huge an investment? Facebook invested USD10 billion in its metaverse division in 2021 alone. [12] And they aren't the only tech giant to be taking a billions-of-dollars punt on it. Some predictions even suggest that eventually the technology will become as ubiquitous as the existing internet, and that functioning without it will be bordering on a logistical impossibility.

This is all wonderful and fascinating, but what does this have to do with building a sustainable future, I hear you ask! Well, the Metaverse actually offers big opportunities to deliver incredible experiences that are far more ecologically friendly than the real-world alternative. One such idea is that eventually it will be possible to use this grand example of the force of Adapting to adapt our experience of international travel through near-real-life experiences of some of the world's greatest landmarks. In this case the Metaverse would utilize immersive technologies such as VR and AR so that you never have to leave the comfort of your own home. There are even companies out there Adapting our ability to sense touch and smell in these virtual environments through the development of special multi-sensory exoskeletal suits. Imagine sitting on a remote beach on one of the islands of Fiji, sipping a virtual cocktail you can taste, feeling the sensation of the ocean lapping at your feet, the gentle blowing of a breeze – all without leaving your living room. Are you ready player one?

Considering that it is predicted that 5.3 percent of all carbon emissions will be attributed to tourism by 2030, if the Metaverse was able to participate in reducing those emissions it would be a game changer for our planet. [13]

Other potential game-changing benefits for our planet associated with the Metaverse include its extended ability to substitute real items for virtual goods and services, and to replace other carbon-producing real-world experiences with equally immersive virtual events. Virtual office environments that enable all the benefits of social interaction with your work colleagues – without the need to travel to the office – are already being experimented with in the Metaverse.

The Metaverse also has a significant opportunity to disrupt the decentralization of other governance and economic mechanisms – all of which have the potential to create healthier social discourse as well as the adoption and diffusion of sustainability solutions.

Indeed, with the Metaverse being predicted to contribute USD3 trillion to the global GDP by 2031 it is time to sit up and take notice of this powerful application of the transformational force of Adapting, and work toward consciously crafting a Metaverse that not only serves our financial and cultural needs, but also our sustainable future. [14]

Summary

Through these examples, I hope to have given you a glimpse of the possible applications of Adapting and its potential to revolutionize sustainability and overcome some of the major sustainability barriers in our operating systems. If the force of Adapting starts to scale for sustainability, we will see massive transformational outcomes, including the following:

First, as Adapting continues to transform the consumers relationship with the ownership of products it will open the door to a plethora of entirely new service offerings across almost every sector and industry.

Second, it would mean decentralized governance mechanisms enabling our global tribe to engage with each other and the public and private sectors to make valuable contributions toward the sustainability agenda in a fair and equitable manner. These mechanisms would also allow for fair and equitable access to and distribution of benefits whilst addressing the need for accountability and responsibility too.

Third, it would allow us to cluster like solutions, resources and technologies as well as identify resources, innovations and stakeholders that may be missing which can make the biggest impact in this process, and if made visible to the coalition of the willing, these could result in faster outcomes with respect to innovative discoveries, support and scale of existing solutions as well as enabling widespread adoption from governments, industry and civilization in general.

Lastly, thanks to emergent technologies like the Metaverse, it would allow for new and more sustainable products and services and the methods to engage with them to emerge in an almost frictionless environment ensuring widespread adoption whilst catalyzing further self-organization and emergence, making our systems, among other things, more agile, resilient and regenerative.

And yet, to achieve these transformative outcomes, many enablers are still needed.

First, taxonomies (the classification of data and information) and ontologies (the links between these classifications) for systematic interoperable solutions are vital, as they can play a major role in ensuring interoperability between technologies, sectors, industries and operating systems.

Public-private partnerships will also be increasingly important in taking climate action, and it is therefore crucial that safeguards are put in place so that all stakeholders benefit from them. Diversity and inclusion are important concepts to be nested into these considerations. This also applies to the future of open-source communities, which perform better both technically and economically when diverse perspectives, experiences and skill sets are included.

Finally, similarly inclusive processes and public participation in governance framework solutions are certainly advisable, not only because this promotes inclusivity, but also because better decisions and outcomes are likely when decision-makers are offered access to more complete information, facts, values and public perspectives. A digital government can also be a major enabler for numerous stakeholders, including citizens, consumers, businesses and governments, offering context that is highly personalized. Citizens with particular areas of interest and even expertise can join together in providing both experience and insight, in the process helping to communicate their views on government and playing a role in creating better services as well.

Yes, there is a huge amount of information to take on board in this chapter, but perhaps the most apt conclusion is a simple one. The world is changing on multiple levels at a pace that has never been seen previously, making harnessing the transformational force of Adapting critical to our future journey.

References

1. Bologna, M., & Aquino, G. (2020). "Deforestation and world population sustainability: a quantitative analysis." *Sci. Rep.* 10: 7631.
2. Global Market Insights. (2021). *Car Sharing Market Size…Forecast, 2021 – 2027.* Available at July. www.gminsights.com
3. United Nations Sustainable Development. (2022). *Quantifying the Impact of Car Sharing.* Available at www.sdgs.un.org.
4. Cherowbrier, J. (2022). *Automotive Industry Worldwide – Statistics & Facts.* Available at www.statista.com. Accessed November 17.
5. BBC. (2019). *China Car Sales Fall for the First Time in 20 years.* Available at www.bbc.com. Accessed January 19.
6. International Energy Agency. (2022). *Improving the Sustainability of Passenger and Freight Transport.* Available at www.iea.orgxxx
7. Parker, T. (2021). *How Car Sharing Reduces Carbon Emissions.* Available at www.karshare.com/ . Accessed November
8. Amatuni, L. et al. (2020). "Does car sharing reduce greenhouse gas emissions? Assessing the modal shift and lifetime shift rebound effects from a life cycle perspective." *Journal of Cleaner Production*, 266, September.

9. Archer, G. (2017). *Does Sharing Cars Really Reduce Car Use?* Available at https://transportenvironment.org. Accessed June 17.
10. Zhang, Y. & Mi, Z. (2018). "Environmental benefits of bike sharing: a big data-based analysis." *Appl. Energy* 220: 296–301.
11. Miller, C. (2019). Taiwan Is Making Democracy Work Again. It's Time We Paid Attention. Available at www.wired.co.uk. Accessed November 26.
12. Kastrenakas, J. & Heath, A. (2021). Facebook Is Spending At Least $10 billion This Year on its Metaverse Division." Available at www.theverge.com. Accessed October 25.
13. World Tourism Organization and International Transport Forum (2019). *Transport-related CO_2 Emissions of the Tourism Sector – Modelling Results,* Madrid: UNWTO. Available at www.e-unwto.org
14. Christensen, L. & Robinson, A. (2022). *The Potential Global Economic Impact of the Metaverse.* Analysis Group.

Part Three

When These Forces Converge

9 The dark side of the force

Looking at the forces through the right lens

If we are to truly understand how to harness the Five Forces of Digital Transformation and their related technologies, we need to first address the mindset that we use to approach them.

For example, most people equate the advancement and impact of these forces to the forms of technology they themselves have personal experience with. We think of our experience in the digital realm in terms of how it personally impacts us and our immediate environment.

For many, the lens of our own experience of technology never goes beyond what is already globally accessible and mainstream. This often constrains our understanding of and experience with the five forces to basic tools like the internet, personal computers, mobile phones, gaming consoles and wearable devices. We rarely have any need to think more broadly about these digital forces or their related technologies, nor the infrastructures that underpin our devices or what else they are being used for. Which makes sense. Who really has the need or the time to do a degree in computer science to make sense of it all, right?

Additionally, as we touched on in Part One of this book, humans are hardwired to process change or cause and effect relationships in a linear capacity, typically in two dimensions. We tap on 'X' app on our screens and get 'Y' result. Easy. This kind of attitude toward technology keeps us snuggled up in our comfort zone and allows us to continue to buy-in to the illusion that change only happens in small, incremental steps over a long period of time.

Seems simple enough but when we step away from our own experience of technology and begin to even try to digest the full spectrum of implications of the impact the Five Forces of Digital Transformation could be having on our lives – we quickly realize we need to replace our linear, two-dimensional lens with an exponential, multi-dimensional one. Once this happens, we begin to understand the massive transformational potential that the Five Forces of Digital Transformation offer, especially as they converge with one another.

Suddenly, we would see our personal computers and mobile phones that were so easy to understand when cause and effect were two-dimensional, now

DOI: 10.4324/9781003187523-12

morph into interfaces that offer access to multi-dimensional technologies capable of unlimited impact on our own lives, our economy and the fate of the planet. They are now, when combined with the other 13.1 billion devices presently in the world today, [1] a gateway to complete transformation of all our operating systems.

However, being human means that we only get temporary glimpses through this exponential lens – it is a constant juggling act to override our own cognitive system and its natural tendencies. On the one hand, we become awakened when the exponential lens is on and revel in the extraordinary advancement of our own making. On the other hand, when the impacts of our digital solutions begin to multiply, converge and cascade, we realize its exponential capacity can spin way out of our control. This causes us to move into the territory of unintended consequences and wicked problems. We can no longer process or make sense of the multi-dimensional exponential impact that we have triggered. We become overwhelmed, flip lenses, and revert to our linear mindset focusing on how this new tech impacts things in our immediate control. Deep breaths. Comfort zone re-established. All is well with the world again. Phew!

Even when we are involved in professions that directly expose us to the cutting edge of digital technologies we can rarely peer through the exponential lens long enough to consider the consequences of our creations nor how they will intersect and connect with other core operating systems. We become blind to how the technology we build interacts with other technologies outside of our intended purpose or 'use case'. We therefore misdiagnose and underestimate the impacts that will emerge when it is put into an environment where it can converge with all the other technologies in the digital ecosystem.

The result? We miss the bigger picture and severely underestimate the speed and scale such digital technology is capable of. We also forget that our technologies continue to exponentially evolve – with or without our help. Yes, technology is now literally capable of having a mind of its own.

So what if we stopped using a linear mindset? What if we had the courage to put the exponential lens on and made the choice to keep it on? What potential outcomes from Harnessing the Five Forces of Digital Transformation would we see? Would these outcomes be helping or hindering our planet?

The acceleration of acceleration of technology

Perhaps we would start to come to an understanding that we are active participants in this digital transformation on a 365/24/7 basis. The ripple effect of this goes so far beyond our baseline understanding of all things digital – permeating industries, countries, sectors, cultures and economies in ways we have been largely naive about and far too ambivalent toward.

Perhaps we would also become sensitive to the true size and scale at which these forces are changing the context of our life experience and the construct under which we create it. As per the example used in the section above, we would no longer look at personal computers and mobile phones as a modern-day marvel but would see them as interfaces to a vast set of digital data, technologies, standards and infrastructures.

As Peter Diamandis and Steven Kotler argue, "*accelerating individual technologies are converging with other accelerating individual technologies, producing overlapping waves of change that threaten to wash away almost everything in their path*". [2] In short – the acceleration of acceleration of technology. So, in viewing the transformational potential of the five forces – through an exponential lens – it is safe to say that there is no going back to a pre-digital era for us. The genie is out of the bottle. The toothpaste can't be put back in the tube. The same will stand for our approach as to how we go about using these forces for sustainability's sake.

Indeed, how we use these forces as part of our global strategy for sustainability as well as how we govern them and integrate them into the fabric of our operating systems will constitute some of the most critical decision-making humanity will make in the next decade. [3]

And so, while I have spent almost an entire book conveying my perspective on the potential benefits of the Five Forces of Digital Transformation, I am far from being a myopic tech-optimist. These forces, and their respective technologies, carry many inherent risks and unintended consequences, simply because they cannot be separated from their human operators. They do not exist in a social vacuum. Every technology has a risk and sustainability profile that can be used to either help or harm. For example, drones can be used to plant trees and also launch weapons of mass destruction, crispr gene editing can be used to increase the drought tolerance of staple crops, and as easily create biohazards that can destroy them.

It turns out that the unknown unknowns about these forces, in terms of sustainable development, are critically important, and have dramatic unforeseen consequences. We must understand that from these transformational forces, digital technologies have emerged, most of which carry such high-risk profiles and associated impact probabilities, that a high level of awareness and preventive action is needed to ensure that we mitigate the risks of the technologies we create *before* their diffusion into civilization begins. The polarizing effect of social media, the attention economy, the inbuilt gender bias in AI, and surveillance capitalism are all examples of unintended consequences of the Five Forces of Digital Transformation with global political and sustainability implications. Yet these innovations were released into the world as part of a grand social experiment, with virtually no safeguards. Meaning all the benefits of digitalization can also be risks when the unsustainable ideas, bad-faith discourse, conflicts of interest and misinformation coded in these technologies also begin to connect, accelerate and scale.

If you've still managed to keep those exponential goggles on, you may also begin to grasp that we are now living in a world where our digital technology is moving faster than our present ability to govern it. Our institutions at all levels are stuck in the rut of a linear mindset, and therefore cannot keep pace with the speed and scale of technological innovation or the transformational forces driving it.

Recall the definition in Chapter Three given to transformational forces of a digital nature:

speed + scale + **purpose** = transformational change

In my view, it is now our collective responsibility to ensure the Five Forces of Digital Transformation are used for the **purpose** of creating a sustainable existence for our people and planet instead of being wielded for the **purpose** of business as usual which happens, by default, to entrench or amplify existing patterns of inequity, marginalization and unsustainability.

Unfortunately, actors who wish to continue to support the business-as-usual narrative are not erecting neon signs anywhere to point us in the direction of how they are utilizing the Five Forces for Digital Transformation to amplify, speed, scale and perpetuate their agenda. Additionally, due to the exponential impact of these forces, actors don't even need to purposefully set out to do harm. All they need to do is develop a new product, service or technology that utilizes one or more of the forces, without specifically engineering it to have the best of our human characteristics and values of sustainability at its core, to unwittingly release potentially harmful exponential impacts on the world.

That's why we need to continually remind ourselves that our penchant for the linear lens means we are often going to be surprised by the catastrophic unforeseen consequences of our creation.

Indeed, as much as I am optimistic about tapping into the Five Forces of Digital Transformation and the digital technologies behind them, I am equally concerned about a future where the very same forces are misused and abused. The unsustainability of our current linear economic systems will only be amplified and accelerated when existing business models further leverage, for the **purpose** of profit, the exponential growth potential of these forces.

Additionally, growth in the use of digital technologies is causing harm to the environment from technology-driven hyper consumption, increased energy demand, material supply chains, and the disposal of e-waste. There are also major governance issues to resolve, involving individual privacy and data security, surveillance capitalism, algorithm bias, fake news and misinformation, the prevention of digital colonialism, and the growing risk of monopolies and digital power asymmetries.

I could write multiple books on the risks of digitalization and the unintended consequences of the Five Forces of Digital Transformation as they converge to impact on and accelerate because of each other. In fact, many have. And even though I do not have the necessary number of pages left in this book to add the weight to this topic that I feel it deserves, I do think it is important that we at least address some of the risks and unintended consequences that are most critical when considering implementing these forces and their associated technologies for sustainability outcomes. And please, dear readers, while the risks and unintended consequences we unpack together in the rest of this chapter may make you feel a little overwhelmed, please try and keep your exponential goggles on, as our awareness of the dark side of these forces is just too important for us to understand to be burying our heads in the proverbial linear mindset sand.

Are we incentivized to change the world for the better… *really*?

As we've discussed this at length in previous chapters, the digital transition that has been taking place at an exponential rate over the last ten years is fundamentally transforming the nature of capitalism, global economic markets, and social interactions as well as human behaviors and mindsets. Many new characteristics have emerged from the digitalization of the economy that now shape and drive a new form of data-driven market and digitalized capitalism.

First and foremost, thanks to the transformational forces of Socializing, Sense-Making and Embedding, we now live in a digitally networked world where all the actors in the economy have become directly connected through the internet and mobile technologies. This has massively expanded the number of connections within and between each of the actors in the economy allowing massive social networking and, thanks in addition to the transformational force of Valuing, decentralized transactions. This level of hyper-connectivity means that we are more capable of peer-to-peer sharing, accelerated social learning and, thanks to the transformational force of Sense-Making, more collective intelligence than ever before. We also have more capacity to self-organize and adapt to changing conditions and opportunities; insert the transformational force of Adapting here. But this isn't all these forces have done since they've been converging.

At the same time, there has been an exponential increase in the level of political discourse that has shifted into the world of social media platforms. They have almost become essential public infrastructure – much like utilities – but not governed in a similar manner. This has been a double-edged sword. While people have been able to connect and share information and ideas on a national and even global scale, they have also been able to amplify misinformation as well as conduct dialogue without accountability. This has often led to toxic exchanges, intimidation, narcissism, hate and filter bubbles. Social media business models have benefited from the amplification of mis-information and division. Ideas are often rated and amplified not based on their quality, but on the amount of jealousy, outrage or polarization they can spark. Trolling became a digital blood sport due to the complete anonymity offered by the platforms. Freedom of speech somehow also became freedom of reach even irrespective of objective facts. This has also brought about the emergence of a phenomenon now known as the 'cancel culture'. Which basic-ally means, if you are voicing an opinion a platform doesn't want you to have, you will be effectively 'canceled' from engaging in educated debate or with your audience on that platform. Game over for diversity of thought at scale.

As most economic activity and social transactions are now computer-mediated, there has also been a proliferation of information about our individual online interactions, behaviors and preferences. As noted by Michal Kosinski, Associate Professor in Organizational Behavior at Stanford University Graduate School of Business, "*All of our interactions are being mediated through digital products and services which basically means everything is being recorded*". Extracting insights from this 'data exhaust' became a new service and business model leading to the growth of the most powerful companies in the world. This data exhaust is increasingly being mined and used to influence our behavior and human agency. Through such use cases of the Five Forces of Digital Transformation this now enables us to create such an accurate picture of who you are, and what target demographic you belong to that, when put in the hands of tech giants, such as Facebook, they are even able to predict things such as knowing when you are pregnant – often before you have told your partner![4] A little (okay, A LOT) creepy don't you think?!

The name of the game from the corporate perspective isn't about making it easier for you to know yourself and have what you need at your fingertips. No. In the context of business-as-usual, it is all about increasing the odds that you will buy from those that are filtering you, that you will believe them or even change your vote for them, and that you will influence others for them in the process too.

One of the scariest things about any digital platform that uses the transformational forces to help understand and shape our preferences, relates to the incentives that drive its underlying business model. Maximizing revenue and influencing your preferences is the goal. There is also zero incentive to care about the quality of content, the authenticity of the source of that content, or any underlying sustainability considerations of a product, service or organization that is responsible for them.

These mechanisms and their algorithms are so subtle in their ability to seduce us into accepting their way of thinking, doing, buying, and so on, that we can no longer be certain of what percentage of our preferences, beliefs or even our human agency, are really our own anymore. This is a critically important point with respect to our ability to achieve sustainability outcomes. If you have time, you may like to read this paragraph again from the top, to let the gravity of this fully sink in.

Despite this power, such technologies have been enjoying a spell in the wild west. Almost no rules or regulations are in place as to how the Five Forces of Digital Transformation we explore together in this book, are used. And because the technologies relevant to these forces are most often embedded in the backend of the interfaces that we use, it has remained largely ungoverned, invisible to the majority, and left to the whims of the media, corporate profitability goals and government agendas. There is little public discussion or concern about the consequences it causes, or about the sustainability opportunities that it could help drive forward and transform.

You would have needed to be living under a rock to have missed the Facebook–Cambridge Analytica data breach which occurred when the personal data of 50 million Facebook users was harvested without users consent by Cambridge Analytica. [5] This hoard of data was then used to manipulate the political mindset of voters leading up to the US election through micro-targeted content driven by AI-based psychographic profiling. Nearly a quarter of potential US voters were put at risk.

When the scandal broke it sparked a US Congressional hearing as well as a global online movement, #DeleteFacebook, which trended on Twitter. The scandal revealed another dark side of the algorithms driving the attention economy and even the very nature of our democratic processes. Author Douglas Rushkoff launched a scathing criticism that has begun to resonate: *"People are at best an asset to be exploited, and at worst a cost to be endured. Everything is optimized for capital, until it runs out of world to consume."*[6] His main idea is that our platform algorithms and digital business models are not optimizing for human happiness, healthy social discourse, and the best of human-centric and sustainability values. No, it is profit-driven decisions that are being made to design these algorithms to shape our collective values in ways that do not serve us well. This must stop for us to have any hope of a healthy civilization, the by-product of which is a healthy planet.

Indeed, technology is now the hand that is rocking the cradle, with its grip on our world squeezing tighter every day. As more choice is made available to us, and business as usual becomes smarter in how it manipulates the Five Forces of Digital Transformation, to craft the decisions it wants to make for us, it will become virtually impossible to find the data, ideas, people, and assets that are truly in the best interests of our individual selves, our civilization and the planet – unless we start doing something about it now.

The dilution of human experience, culture and connection

It isn't only our ability to think and act that is being hijacked by the way business as usual is presently harnessing the collective power of the Five Forces of Digital Transformation. It is also our ability to connect with lived experience and culture, arguably the very essence of our humanness.

In a panel conversation on the possibility of building a digital planet for sustainability, Caleb Behn, a Legal Policy Advisor to the National Chief Assembly of First Nations (Canada) asked the question – *"how can we connect data to the human spirit and how do we deal with things that can't be measured"*? He followed on to ask *"If words don't have a meaning without context, how can we understand and give meaning to data without context?"* These are critical questions and are a potentially huge blind spot in the business-as-usual agenda.

Not only do data and digital insights only measure part of a very complex picture – but they might undermine our direct experience of an issue. Can you have a profound feeling for a specific place if you only experience it in a digital form? There is a risk that if we rely too much on how these forces are presently being applied to our lives, we may miss out on some of the most essential aspects of being human, which cannot be measured, such as the state of awe or the ability to navigate ambiguity and paradox.

A great example of this phenomenon was revealed in the documentary "Chasing coral". The filmmakers had set out to document a massive coral bleaching event in the Great Barrier Reef using satellite analysis and a series of automated underwater time lapse cameras. The filmmakers admit that they designed the project to run without emotion – they were documenting the event through a data-driven lens and through the digitalization of reality. But were they connecting to the lived experience of it?

It wasn't until the automatic cameras failed and they were forced to do manual daily photographs of the dying coral that they began to experience the bleaching at an emotional level. Zack Rago sums up the heart wrenching experience as "difficult". And perhaps this is the take-away – be aware that these forces are capable of shielding us from direct experiences – stripping us of a deep emotional connection to a specific issue, person or place. We must remember 1s and 0s cannot capture everything.

Author Douglas Rushkoff notes "*True awe is timeless, limitless and without division. It suggests there is a unifying whole to which we all belong – if only we could hold on to that awareness*". [7] We may also fail to fully consider other ways of seeing, understanding and connecting to the world held by different cultures and indigenous groups. These range from the dreamtime, dream space, sky camp and spirit world concepts of indigenous groups, to the Buddhist belief in karma, to the ancient Chinese philosophy of yin and yang. Given our inability to think beyond our linear mindset, how can we possibly know what the long-term consequences would be to our social and ecological systems and civilization, if these ancient wisdoms were left behind because the lived experience of them didn't fit into business-as-usual's profit-centric digital design?

How are these ways of knowing our world being included or excluded in the design of our digital technologies? How can we use the Five Forces of Digital Transformation to preserve the ancient ways and languages of our indigenous tribes, ensuring our civilization can benefit from them indefinitely?

The other critical aspect that must be considered is the way humans go about trusting others and building long-term relationships. Enter another neurochemical – oxytocin – also known as the Cuddle Hormone or the Trust Hormone. It is the "social glue" that adheres families, communities, and societies, and enables us to engage in all sorts of interactions and transactions. [8]

There is a debate raging in the scientific community on whether digital encounters with other humans can generate the same depth of emotional connection, trust and release of hormones as do physical encounters. In other words, does the digital space trigger the same intensity or same chemical composition – or only lead to the shadow of an experience?

It turns out that social media can trigger the release of oxytocin as well as other neurotransmitters such as serotonin and dopamine. This could have massive ramifications for building a global level of empathy with fellow humans that we never directly meet. A global coalition of the willing for sustainability.

However, an interesting paradox has emerged. While we are more digitally connected to each other than ever before, levels of anxiety, depression, loneliness and isolation have never been higher. [9] For example, people who report spending more than two hours a day on social media [10] had twice the odds of perceived social isolation compared to those who said they spent a half hour per day or less on those sites. Similarly, people who visited social media platforms 58 times or more per week or more, had more than three times the odds of perceived social isolation as those who visited fewer than nine times per week. Is social media the world's first digital drug? How matrix.

The conclusion is clear: *"we have evidence that replacing your real-world relationships with social media use is detrimental to your well-being"*. [11] This goes hand in hand with the phenomenon of the 'online disinhibition effect' – the idea that people behave online in ways they typically would not when in-person due to the technological separation. While it is too early to say how much an entire global population experiencing symptoms of depression and anxiety is due to technology use, or will be impacted in their ability to address their behaviors and decisions around sustainability, we do know that climate change itself is further compounding the issue of mental health, so much so that the IPCC released a report stating that people suffering from an existing mental illness are three to six times more likely to die in a heat wave. [12] Can the conclusion then be comfortably made that the second order consequence of social media and our addiction to devices making more of our population mentally ill, is that even more people will die as temperatures rise due to the impacts of climate change ravaging our planet? If we do not keep our exponential goggles on, we run the risk of having to find out the answer to questions like this the hard way.

Even if the possibility of this being the case is small, should we not consider, before we run the largest experiment in the history of humanity in terms of digital social relationships and brain chemistry modification, that perhaps we should understand the short and long-term consequences from more perspectives? Surely a few precautionary principles couldn't hurt? And should we ask ourselves whether putting the 'master switch' in the hands of a few companies with more financial power than many countries is wise?

At the very least, the impacts of the consequences listed in this section should make us think deeply about how, future forward, we can ensure the ways in which we use the Five Forces of Digital Transformation are entrenched with the right purpose and values at their core.

All the upside minus the accountability

When the likes of AirBnB, Uber and Kickstarter sparked the share economy, they didn't just create a playing field where anyone could come in at entry-level and own a piece of an industry's market share. No. By catalyzing the diffusion of profits and market share, they also unwittingly opened the doorway for the diffusion of power, responsibility, agency and action within those industries.

More and more industries are undergoing severe metamorphoses as the reach of the share economy pervades them, and with it the power and influence of large enterprise and heavily regulated processes also begins to diminish proportionately. Indeed, the rise of the sharing economy is anticipated to be so rapid that PwC data suggests that companies that form the sharing economy will be responsible for 50 percent of the market across five specific sectors by 2025. These areas include, among others, media, transport and accommodation, and would represent a tenfold increase in the market share of sharing economy companies in just 12 years, and an annual revenue rise in the region of USD320 billion compared to the same figure in 2013. [13]

Let's consider what the full weight of an economy built on models such as the share economy has on efforts to mitigate climate change.

Right now, to remain in business, hotels need to have a registered business entity, and undergo rigorous regulatory approvals, as defined by their local and federal governments. If they don't adhere to certain standards and regulations, they can't keep their doors open for business on an ongoing basis. Now, let's say, hypothetically speaking, that in 2025 local and federal laws for the hospitality and tourism industry change, so that accommodation providers, such as hotels, now need to adhere to strict net-zero carbon emission guidelines.

Yet the total sum of carbon emissions for the hospitality and tourism industry, by 2025, isn't just owned by established hotels and other actors in the private sector. PwC estimates at least 50 percent of it will be owned by the John and Jenny's of this world, who are participating in the share economy to have equal and fair rights to the same marketplace, by renting out their small apartment. This is where the situation can get very tricky for sustainability. John and Jenny just have a small flat, and only rent it out through AirBnB on an ad hoc basis. Thus, they don't fall under the same regulatory frameworks or reporting conditions or performance criteria as a hotel, and, therefore, don't need to adhere to the same net-zero carbon emissions guidelines as the hotels would have to. Yet, there are now millions of AirBnb hosts just like John and Jenny all over the world. In fact, there are so many that they now make up 50 percent of the market share of this economy. The total of the revenue, market share, buyers' trust, and so forth, that the John and Jenny's will have, in the very near future, will be equal to, or more than, the major industry actors from the hospitality and tourism industry. Say what?! Yes, you read that correctly.

It then stands to reason that the impact on sustainability made by the John and Jenny's of the share economy has the potential to eventually be greater than that of **ALL** the traditional industry actors combined.

In short, we can no longer rely upon the power brokers, influencers or monolithic corporations to take full responsibility for implementing sustainability solutions to mitigate climate change. Their impact simply won't be great enough any longer, as we've already diffused far too much of their wealth, and with it their power, influence, and decision-making capacity. Generally, as we discussed in the previous chapter with respect to digital governments, this is a positive phenomenon, but in terms of the ecology of the planet, we also need to understand the socio-economic consequences.

We are now, for the first time in human history and as a direct result of the exponential impact of the transformational forces that have catalyzed the share economy and the decentralization of governance and economic mechanisms, in a position where the negative impacts on sustainability caused by industry now need to be acknowledged, taken responsibility for, and acted upon by the entirety of our global tribe, based on the proportion each individual actor contributes to that which was once only attributed to the major industry players.

Furthermore, as you are now well aware, this is but one example of how the transformational forces can cause paradigm shifts in entire industries. In releasing and propelling the very technologies that we needed to create equal, just and accessible verticals within our economy, we have also unwittingly handed ourselves the ability to catalyze and scale – at warp speed – our very own destruction. Indeed, in common with every other risk highlighted in this chapter, depending on what incentives, purpose and values are used at their core, the Five Forces of Digital Transformation provide both our opportunity and our undoing.

The rise of the power platform and the divide left in its wake

The global data economy is now valued at €3 trillion and personalized data is valued at €1 trillion. [14] Data became the world's most valuable commodity, and it is the new raw material playing a key role in value creation. Apple, Amazon, Alphabet and Microsoft are now worth USD5 trillion, the same as the combined GDP of 135 countries or the entire London stock exchange. [15] The world of fintech has also exploded where financial apps and related technologies are completely revolutionizing the flow of money around the economy and through peer-to-peer transfer.

Due to the explosion of information, many new tools were needed to handle this structural change. A very narrow range of profit-maximizing values are now being unleashed and imposed on the entire world. A new form of digital colonialism is emerging where tech companies have expanded their products and values across the globe, massively undercutting our cultural diversity and national sovereignty. They extract data and profit from users all around the world while largely concentrating power and resources in two countries: The US and China. [16]

Some global digital platforms have achieved very strong market positions in certain areas. For example, Google has some 90 percent of the market for internet searches.

Facebook accounts for two-thirds of the global social media market and is the top social media platform in more than 90 percent of the world's economies. Amazon boasts an almost 40 percent share of the world's online retail activity, and its Amazon Web Services accounts for a similar share of the global cloud infrastructure services market. In China, WeChat (owned by Tencent) has more than one billion active users, and, together with Alipay (Alibaba), its payment solution has captured virtually the entire Chinese market for mobile payments. Meanwhile, Alibaba has been estimated to have close to 60 percent of the Chinese eCommerce market. While these titans of technology haven't yet hit total monopoly status, they have become as important as public utilities to the modern world – but often operate with few safeguards to protect the public interest.

Fewer than 20 companies own more than 80 percent of the internet's capacity, which is based on storage and computing. [17] There is no level playing field in terms of giving sufficient space for other technology solutions to emerge that might reflect sustainability-centric values in their algorithms. If they do, they tend to be squashed or bought out by the major firms to avoid any potential competition.

Big Tech firms have effectively become gatekeepers to the digital economy, with the power to pick winners and losers, shake down small businesses, and enrich themselves, while choking off competitors. Author Scott Galloway observes that "The Big Four" major technology companies (Amazon, Apple, Facebook and Google) are effectively in a race to run the literal operating system of our lives". [18]

Similarly, while the thriving "gig" or "platform" economy promises to free workers from the 9-to-5 work environment, the reality is quite different. Not only do gig workers lack health care, retirement benefits and collective bargaining voice, they are typically at the mercy of their employers' scheduling needs and the relentless apps that task them.

While the gig economy is a critical onboarding pathway for employment, it also profits off desperation, servitude and predatory relationships. The gig economy has created an employment model that often strips workers of the rights they've earned through more than a century of fighting for them. It uses automation not to make a better world for everyone, but to drive down the cost of labor while outsourcing the costs and risks of physical assets to their workers. Are these uses of technology optimizing for the right set of social outcomes or simply magnifying the divide between the served and the servants? [19]

Indeed, the WEF Global Risks report for 2021 listed "Digital power concentration" and "digital inequality" as numbers 6 and 7 on the critical short-term threat list – both representing a clear and present danger to social and political stability. [20] Concurrently, four of the five top risks on the WEF index were environmental: extreme weather (rank 1), climate action failure (rank 2), human environmental damage (rank 3) and biodiversity loss (rank 5). The connection between a healthy civilization and a healthy planet is blatantly obvious.

The further our technology advances, the greater the digital divide becomes. Half the globe's population is being left behind. Can we really afford so much of humanity out of action in the race against climate change and in facing our other global challenges? How can we be sure of the on-ground impact of climate change if half the population is unable to stream high quality in-situ data from their geolocations? It isn't just connectivity that these populations lack, it is infrastructure, knowledge, access to data, tools and so much more.

People affected by this digital divide are overwhelmingly from groups who are already marginalized: women, elderly people and those with disabilities; indigenous groups; and those who live in poor, remote or rural areas. Many existing inequalities – in wealth, opportunity, education, and health – all of which have also been highlighted for their direct impact on sustainability – are being amplified. [21]

So, the question is how do we use the Five Forces of Digital Transformation for the purpose of dispersing such power granted to the few and the promotion of equality instead of continuing to amplify the already significant threat they pose when used for the wrong reasons?

Big brother is watching you – what could possibly go wrong?

Though it isn't just about companies conducting global economic surveillance, it is also about the ways that governments are using digital technologies to monitor their citizens or engage in intelligence operations. In 2013, Edward Snowden came forward as a whistleblower about the NSA and its international intelligence partners' secret mass surveillance programs and capabilities. Similar programs were revealed in other countries including the UK, Russia, and more. And many countries are actively engaging in cyber warfare and digital destabilization of our democracies. Indeed, the US Intelligence Community concluded that the Russians had deployed a sophisticated misinformation and influence campaign to disrupt the US election in 2016 by undermining public faith in the democratic process. [22]

China has gone the other way entirely. They make no secret of the fact that every one of their citizens' every move, every click and every transaction are being monitored and scrutinized and are equally as forthcoming about how they intend to use this data to influence their citizens' behavior. [23] Their National Social Credit System was due to be fully operational in 2020 and has been designed to value and engineer better behavior for individuals and companies. This monolithic project uses big data to closely monitor the behavior of individuals and corporations to reward the compliant whilst punishing those they deem disobedient.

China intends launching social credit laws in the coming years and whilst nothing has been officially rolled out for individual citizens, the system and its laws have been implemented to trial their effectiveness in demonstration cities across their nation. The 'deadbeat' map is one of the functions released under this pilot program which allows companies to be alerted to individuals that owe them money that have come to travel within a 500-meter radius of their office. It even enables citizens to report on irresponsible dog owners who leave their 'doo-doo' behind.

It might sound somewhat benign in such examples but this "Black Mirror" style project has much, much bigger plans in their sights with reports suggesting things such as finding your credit rating impacted if you are found to be socializing with people with opposing political views to those of the government. Ouch. You might even find your health insurance premium increased because big brother doesn't like how much alcohol you buy as part of your grocery order each week. Or that your children are denied entry to their choice of educational institution because they spent too much time in front of screens and not enough time participating in extracurricular activities.

It gets a lot more insidious when we consider the significant impact such a system could have on altering behavior for fear of being labeled 'disobedient'. What could a government manipulate their people into being, doing and thinking if it deploys its surveillance technology in such a way? And if you think China is on its own in deploying AI global surveillance technology to monitor its citizens, think again. More than 175 countries around the world presently use such technology to monitor their citizens to varying degrees, so where do we draw the line? [24]

The darker side of humanity

While many of the above risks and unintended consequences come about through what could largely be debated as legal use cases of the five forces (emphasis on the word 'debate'), we also need to be aware that in every way that the Five Forces of Digital Transformation can be harnessed for business-as-usual **or** sustainability, they can also be used by those with far more sinister plans, many of which have dire consequences for our planet.

I will be the first to admit that I am no expert when it comes to the topic of technology in the context of organized crime (a whole other book for an entirely different author) so we won't be diving into this in too much detail, but I do feel it is important I at least bring it to your awareness.

For example, the forces can be used to fortify an individual's, organization's or government's cyber-security infrastructure and protocols – as easily as they can be reverse engineered to be used to hack into those same systems. The consequences of such crimes for sustainability may not seem obvious from the outset but, with the help of our exponential goggles, we will be able to grasp how enormous they really are. Why? Because these kinds of crimes destroy trust at scale and in a blink of an eye. There are not many transactions or interactions that can occur in our modern-day existence without at least a basic level of mutual trust being exchanged between parties.

Trust in the services we engage with, trust in the credibility of the products we buy, trust in the platforms we use, trust in the organizations that we transact with, trust in the data we rely upon to make decisions, trust in the use of the sensitive information we share, trust in the governments that run our country and most importantly, trust in each other – can all be impaired when such attacks occur. If we cannot rely upon our information to be held securely or for the data we base our decisions on to be accessible and accurate, we will never be willing to engage on a level that enables transformational change. And, if we cannot engage and collaborate with each other at scale in digitally safe environments, there isn't much hope for sustainability solutions to make the impact they need to.

The above example is just one of so many ways organized crime can impact sustainability. Everything from terrorists trying to access national security systems to deploy nuclear warheads, poachers using drone, geothermal and geo-spatial technology to locate and hunt endangered species, the prolific use of social media platforms for the trade of exotic and endangered animals, the hacking of biosecurity facilities to gain access to biological weapons of mass destruction, the planting of fake satellite images to act as decoys for illegal deforestation operations – even the stealing of carbon credits from national databases to be sold on the black market – and so, so many more. [25] It's not a pretty picture by any stretch of the imagination.

This means we also need to be mindful that criminal intent and its subsequent outcomes are also capable of scale, speed and impact when the Five Forces of Digital Transformation are used for darker purposes.

Thinking about how the solutions we develop can be used in reverse to achieve the opposite outcome of what we set out to do – should this not also be a priority consideration before we release these technologies into the proverbial flywheel?

With great power comes great responsibility

By now, it should be extremely clear to you that we cannot afford to continue to use god-like technology such as the Five Forces of Digital Transformation and manipulation of human agency at a global scale, to advance profit-centric interests to the exclusion of planetary needs and healthy social discourse. Unless we figure out how to combat the dark side of these forces humanity will never have the level of collective focus, understanding, agency and wisdom needed to make the right decisions nor to collaborate at a level capable of effecting real-world change.

For some, such abuses and unintended consequences in the application of digital technologies, such as those I have discussed in this chapter, have caused us to try and scramble for the big pause button in the sky. We are starting to reflect on what we have built and the safeguards (or lack thereof) we have constructed. Indeed, a massive pushback has been sparked against the digital revolution. This 'techlash' is a growing public animosity toward large Silicon Valley platform technology companies and their Chinese equivalents. And how can we blame those who feel this way?

To date, these forces have been used by business-as-usual to amplify the worst of our human impulses and culture of convenience. It is accelerating our level of consumption and reinforcing our linear and unsustainable model of economic growth. But it doesn't have to be this way. Indeed, to achieve a sustainable civilization we must intentionally direct our technology to support better, more sustainable outcomes.

We need widespread and deep reflection and conversation on what we aim to achieve from technological progress and how we can ensure it leads to equity, sustainability and dignity. How can technology help us move from *"evolution by natural selection to evolution by intelligent direction"* ? [26]

Like I felt in writing this chapter, I am sure you have had some very uncomfortable moments throughout your experience of reading it too. Exponential goggles are fun when we are future-casting and painting landscapes together of what could be possible – but less so when we are exploring the negatives. Sorry, not so sorry. You needed to know.

However, despite what we have unpacked in this chapter, I still strongly believe the Five Forces of Digital Transformation pose an incredible opportunity (perhaps the *only* opportunity) to embed the best of humanity and sustainability values into all aspects of our world and efforts as stewards of this planet. Our challenge now – or rather our responsibility – is to move at a greater velocity to encode these more positive outcomes in the architecture of our digital landscape than business-as-usual, or organized crime, can move in opposition to them.

So … are you up for the challenge?

References

1. Nick.G. (2022). *How Many IoT Devices Are There in 2022? [All You Need To Know]*. Available at www.techjury.com. Accessed November 26.
2. Diamandis, P. & Kotler, S. (2020). *The Future Is Faster Than You Think: How Converging Technologies Are Transforming Business, Industries, and Our Lives*. Simon & Schuster.
3. Ibid,
4. Haiek, C. (2019). *Facebook Might Know You're Pregnant Before You Even Tell Your Partner*. Available at https://news.com.au
5. Cadwalladr, C. (2018). *Revealed: 50 million Facebook Profiles Harvested for Cambridge Analytica in Major Data Breach*. Available at https://theguardian.com. Accessed March 17.
6. Rushkoff, D. (2021). *Team Human*. Norton & Co.
7. Ibid.
8. Penenbeg, A.L. (2010). *Social Networking Affects Brains Like Falling in Love*. Available at www.fastcompany.com. Accessed January 7.

9. Karim, F. et al. (2020). "Social media use and its connection to mental health: A systematic review." *Cureus*. June, 12: 6.

10. Hobson, K. (2017). *Feeling Lonely? Too Much Time On Social Media May Be Why*. NPR.

11. Ibid.

12. Hansen, A. et al. (2008). "The effect of heat waves on mental health in a temperate Australian city." *Environmental Health Perspectives* 116: 10, 1369–1375.

13. Robinson, D. (2019). *Sharing Economy Companies List: From Airbnb and Uber to Omni and Hubble, 15 companies disrupting their industries*. Available at www.ns-businesshub.com. Accessed March 25.

14. Thirani, V. (2017). *The Value of Data*. World Economic Forum.

15. Feng, E. (2020). *How Apple, Amazon, Alphabet, and Microsoft Became $1 Trillion Companies*. Available at www.marker.medium.com. Accessed March 3.

16. Kwet, M. (2018). "Digital colonialism: US empire and the new imperialism in the Global South." *Race and Class*. 60(4).

17. Koetsier, J. (2020). *'Largest Distributed Peer-To-Peer Grid' On the Planet Laying Foundation for A Decentralized Internet*. Available at www.forbes.com. Accessed June 20.

18. Galloway, S. (2017). *The Four: The Hidden DNA of Amazon, Apple, Facebook and Google*. Bantam Press.

19. Marx, P. (2016). *How the Gig Economy Profits Off of Desperation*. Available at https://thebolditalic.com. Accessed October 12

20. World Economic Forum. (2021). *The Global Risks Report 2021: 16th Edition*. Available at www.weforum.org

21. United Nations. (2021). *Secretary-General's Roadmap for Digital Cooperation*. Available at www.un.org

22. Miller, C. (2018). *The Death of the Gods: The New Global Power Grab*. William Heinemann.

23. Zhou, C. & Xiao, B. (2020). *China's Social Credit System is pegged to be fully operational by 2020 – but what will it look like?* Available at www.abc.net.au. Accessed January 1

24. Carnegie Endowment for International Peace. (2022). *AI Global Surveillance Technology*. Available at https:www.carnegieendowment.org

25. Williams, A. (2022). *Are You a Victim of Stolen Carbon Credits?* Holman Fenwick Willan.

10 Convergence by design

Looking at the operating systems with our exponential goggles

In our previous chapter we began using our exponential goggles to understand the potential risks and unintended consequences of the Five Forces of Digital Transformation. Yet, these goggles can be used for SO much more.

Let's explore what happens, when we look at our operating systems, discussed at length in Chapter Two, through this new lens.

When the goggles go on we begin to not only fully appreciate the bugs, perverse incentives, brakes and accelerators in each of the individual operating systems, but also how these systems can connect and interoperate as one complete whole. I like to refer to this as the 'whole-systems' perspective.

Indeed, this kind of "systems integration" thinking, as is also championed by other notable authors such as Isabel Rimanoczy, and is the kind of lens we need to look through if we want to have any hope of using the Five Forces of Digital Transformation to achieve a sustainable civilization. [1] Only then will we be able to design and deploy sustainability solutions at the speed and scale needed to solve our planetary crises in the next ten years.

Without thinking about transformational change at a 'whole- systems' level, we are simply back to the kind of linear or 'business as usual' thinking that got us into this mess in the first place. As our economic markets grow, and the digital revolution continues to evolve, sustainability solutions that do not consider how they will interconnect, interoperate, cascade with, and be amplified across all of these operating systems will create a new combination of system-level bugs and perverse incentives, barriers and breaks, along with resistance from the entrenched interests that they threaten. Additionally, they will also likely cause a further sea of unintended consequences that may undermine their adoption or effectiveness – or worse, exacerbate our global crises.

The easiest way to visualize the interconnectivity of our four human operating systems when operating as a larger "whole-system" is as an infinity loop (see Figure 1.1).

Presently, we have a situation where a solution in one system has unanticipated impacts in other systems that often have significant consequences. As

DOI: 10.4324/9781003187523-13

Human
Cognition

Economic
Incentives

Social Incentives

Governance, Policy
& Regulation

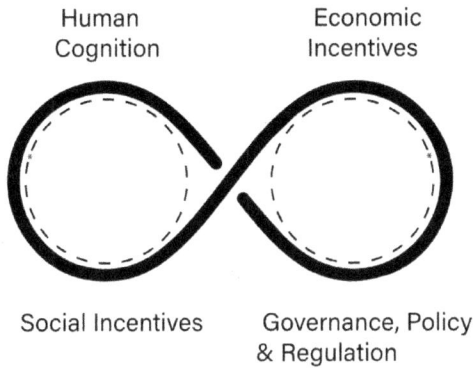

* dotted lines represent the negative feedback loops that are
amplified between individual systems when whole-system failure
points are not addressed

Figure 10.1 Depicts the connection between the four core operating systems and can
be visualized as an infinity loop.

an example, social polarization and the amplification of misinformation and 'alternative facts' are a by-product of the transformational forces causing unexpected interactions between each of the operating systems due to social media applications like Twitter and Facebook.

While a sustainability solution might initially focus on solving a specific problem in one of the systems, it is essential to be thinking about how it can also be designed and deployed so that it evolves and adapts as a multi-system solution to ensure any future interoperability, and therefore impact, across all operating systems, is positive.

I know this sounds complicated and even a little beyond what our individual cognitive abilities might allow – but I believe the greatest opportunity the Five Forces of Digital Transformation can bring to sustainability is in playing this transformative role at the 'whole system' level.

In fact, I see major opportunities where the convergence of the Five Forces of Digital Transformation will help drive four major 'whole system' outcomes in a manner that was not before possible.

1. A common digital language

Digital technologies can improve the connectivity and interoperability across these operating systems by offering a common digital 'language' that can be used to share information and generate feedback loops.

Each of these systems is fundamentally shaped by the manner that information flows within the system and is shared with other systems. The time lags and type of feedback loops that are generated within and across these operating systems ultimately shape incentive structures and response behaviors.

These feedback loops depend on the rate, source, quality, quantity, interoperability and transparency of information. However, the mechanisms for allowing information to flow within and across these systems are massively different in each case.

As an illustration of this, consider the basis for some of the complex systems that we understand and encounter. Human cognition relies on electrical impulses, chemical reactions, hormones and lived experience; social systems rely on language, culture, community and meaning; economic systems rely on cost, price, supply and demand; while governance systems have historically been expressed through laws, regulation and contracts.

When we harness the Five Forces of Digital Transformation to converge by design, we can build the functionality into our solutions to enable interconnectedness of all four operating systems because the 1s and 0s of the digital nature of these forces offer a common digital 'language'. This creates the 'connective tissue' that can be used to share information in digital form no matter which operating system it originated in or how many it ends up interacting with. This digital information can, as a result of these forces converging, flow across these systems and be shared, processed, and acted upon at the speed of light, while incurring virtually no cost. Critically, we can address time lags in feedback loops, or build entirely new feedback loops, so that the correct information is available in a timely manner to support sustainable decision-making and outcomes in the best interests of our people and planet.

Some technology authors, such as Douglas Rushkoff, rightly point out the potential risks of compressing all meaning and value into binary 1s and 0s. [2] This potentially undermines our ability to properly 'code' qualitative concepts, such as trust, meaning, empathy, compassion, love, courage, spirituality and ambiguity.

While I agree with this limitation, I cannot see a pathway to scale global sustainability in the next ten years without a basic common language between these core operating systems – even with these risks and imperfections. So long as we understand these limitations, it may be possible to find ways to mitigate them.

2. A common clock rate

Even if the five forces enable our operating systems to more easily share information and gain access to important feedback loops, we must understand that there are significant disconnects between how each operating system scales, their adaptive and operational capacities, and processing speeds. Former Google Design Ethicist, Tristan Harris, from the documentary *The Social Dilemma* calls these fundamental differences *"varying clock rates and operating frequencies"*. [3]

This means that there is a massive gap between the speed at which each operating system can process information, adapt to change, adopt new behaviors, and evolve. As an example, it takes all of five seconds to write a line of code in a piece of digital technology, but, on average, minutes for markets to react to events, years to adopt laws, decades to change social norms, and millennia to evolve DNA. Powered by the transformational forces, some technologies can also scale at exponential rates at a global level, while many governance solutions take years to implement, and are often constrained by national borders, along with other socio-economic, geopolitical and legal factors.

These fundamental disconnects in the clock rate of each system have also meant that some systems interact to amplify one another, more so than others. I have illustrated these convergent accelerations between such systems by using dotted circles in Figure 1.1.

A good example of this phenomenon is the convergence and cross-amplification between our human cognitive and economic systems. The use of the five forces and their associated technologies, such as AI, has been amplified by the economic system, due to the high return on investment that they offer. These forces have also found ways to 'hack' the human cognitive system – creating digital addictions and dependencies that further accelerate demand and therefore further interact with the economic system. However, unless we can find ways for our governance and social systems to keep up, they will always lag behind technological innovation and the use cases that prevail. This means that they won't be able to address the dark side of these forces before they are unleashed; instead they will always be in response mode, rather than being preventive and adaptive.

The Five Forces of Digital Transformation can be used to help level the playing field between these operating systems by offering a common 'clock rate' to process digital information, as well as enhancing the adaptation, evolution, responses and response times of these systems.

3. *Integration with natural systems*

The third major opportunity the transformational forces have in successfully enabling 'whole-system' interoperability speaks to how our operating systems integrate with our natural systems.

Governance systems are a critical component of our infinity loop diagram. This is because they are the glue that has enabled capitalism to establish connectivity between all our ways of being and doing in this world. While their mechanisms are far from perfect, governance systems support the flow and regulation of behavior, data, information, money, decision-making and transactions.

With governance systems as a core component of our infinity loop, the four core operating systems may have historically worked to manage the human world to varying degrees of success. However, it is clear they have not worked well in the sustainable management of our planet, including its climate, ecosystems and natural resources. Nor have they really worked for the best interests of humanity in terms of wealth equality, gender equity, and the inclusion in society and economic systems of varying stakeholder needs.

Because of this oversight, our existing operating systems typically connect to our natural world systems in unhealthy ways. In other words, the crossover tends to be unidirectional, and either extractive or damaging, rather than sustainable, circular, or regenerative.

For example, information about the planet or our climate systems is not a variable or data point that typically flows across this infinity loop, nor does it become integrated into decision-making within any of our human operating systems. Or, if it does, the feedback loop is typically too slow to support rapid corrective action.

As a result, this information is currently failing to sufficiently shape incentive structures, prices, financial decisions, social norms and response behaviors. For the most part, information about the health of our planet and vital ecosystems is simply ignored, too fragmented to act upon, or generated too slowly to impact decision-making in real-time. As noted, in many cases environmental impacts from economic production are treated as an externality and excluded from being effectively accounted for in our economic systems. Additionally, the consequences of these environmental impacts on all other operating systems are also not being effectively accounted for.

A case in point – despite all our technology, we still cannot measure global progress against 58 percent of the SDG environmental indicators. [4] Our best technologies are not yet being aligned to measure progress against our global goals. Surprisingly, none of the goals explicitly include targets and indicators to harness digital technologies, nor the force of digital transformation, to accelerate and scale progress.

The fundamental reason for this is that all four systems have been developed without considering how humanity's best interest hinges on ecological health. Indeed, the creation of these systems essentially predated understanding and recognition of the importance of ecology. It is obviously time to change this. The only way to achieve greater consideration for the ecological health of the planet is to integrate data from our natural systems within the 'code' of each operating system, as well as at the whole system level.

Up until recently, it wouldn't have been viable to suggest that data about the status and health of natural systems should be embedded into the code of each operating system. Not least because much of this information was not available at the global level in real-time. However, the Five Forces of Digital Transformation are now removing this limitation, making the embedding of this data more feasible.

Overall, the Five Forces of Digital Transformation are creating a massive opportunity to generate and integrate natural systems data across all human operating systems. Digital technologies now offer the potential to send information on natural systems across each individual operating system, as well as the 'whole system' that could, for the first time in our history, shape a completely new set of sustainability solutions, incentives and behaviors.

However, even if data about our natural systems is integrated into each operating system, this won't necessarily drive sustainability outcomes at the systems level.

We need to make a conscious effort toward directing the digital transformation of each operating system to achieve sustainability, only then can we completely shift our outdated economic paradigms.

4. A planet centered (whole) operating system

However, even if data about our natural systems is integrated into each operating system, this won't necessarily drive sustainability outcomes at the systems level. This is where I see the fourth and final major opportunity for the Five Forces of Digital Transformation to play a role in establishing interoperability between our operating systems on a 'whole-system' level.

To achieve sustainability outcomes at 'whole-system' level, we must ensure that the digital platforms, applications, algorithms and filters that underpin each 'operating system' are designed and optimized to achieve sustainability incentives, behaviors and outcomes rather than simply maximizing profit.

Sustainability can be a core outcome of the digital economy – but only if we build the digital infrastructure, standards, algorithms, filters, platforms and governance framework with this strategic intent.

The Five Forces of Digital Transformation now make it possible for us to shift the focus of our operating systems from human-centric design, outcomes and benefits to that of being planet centered.

In my opinion this then changes how our infinity loop will look and function moving forward, see Figure 1.2.

If we use the Five Forces of Digital Transformation to bring our natural systems and planetary outcomes into the center of our infinity loop we can completely shift our outdated economic paradigms and inadvertently address many of the bugs and pieces of malfunctioning code in our operating systems, individually, and as a whole.

With a planet centered 'whole-systems' approach we can finally create a series of signals and incentives among regulators, investors, producers and consumers forming feedback loops where the expectations and actions of each actor help to determine the expectations and actions of every other actor in a virtuous loop. We can finally build self-enhancing and self-sustaining feedback loops that drive and reinforce planetary sustainability, resilience, circularity and regeneration.

Of course, as previously stated, sustainability isn't the only game in town that needs to be achieved through digital transformation. We must also consider creating a set of emergent properties that address other key building blocks of a sustainable civilization. These include innovation, resilience,

Human   Economic
Cognition   Incentives

Our Planet

Social Incentives   Governance, Policy
                           & Regulation

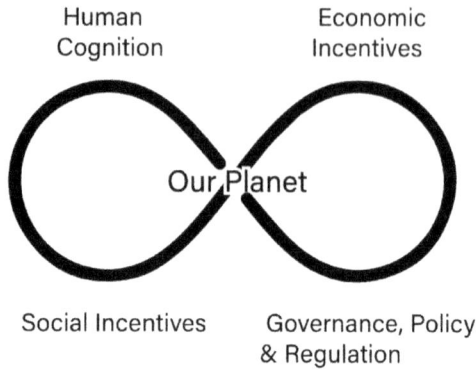

Figure 1.2 Depicts a planet centered, "whole-system" approach by placing our natural systems and planet at the center of the infinity loop. It is proposed this
would also resolve the negative feedback loops as depicted by dotted circles
in Figure 1.1.

agility, circularity and regeneration. We must determine how to design universal capabilities for information flows, feedback loops and incentives so that
our whole system can support these wider emergent properties.

Having a universal means to measure and report on the effect and overall
impact of all sustainable infrastructure systems (SIS), solutions, innovations
and initiatives that are being implemented around the world will be one of the
most important tools in maintaining a data-driven approach to sustainability
and truly understanding how much progress we are making with respect to
restoring health to our planet and in our transformation to a sustainable
civilization.

Standardized reporting enables the global collective to monitor existing
areas of concern, diagnose and triage new problems, test the overall effectiveness of the solutions architecture as well as each solution's ability to
scale and make impact and allows the global collective to remain agile in its
responses to ensure we are capable of iterating and/or pivoting technologies
and new and existing innovations and solutions and their application where
necessary.

It also enables stronger decision-making as we become able to accurately calculate, allocate and manage required resources as well as the kind of
involvement required from each actor and stakeholder group and will be able
to track the impact each resource/actor/stakeholder makes on any given project, among very many other things.

Projects that are not flourishing are also able to be tabled with the collective
intelligence for troubleshooting and are either put on a pathway for success
and scale or sidelined through collective decision before they can waste any

further resources unnecessarily, thus enabling those resources to be diverted to projects showing more potential and or to begin supporting the research, development and launch of new innovations being trialed.

To have a universal reporting system that is capable of such things will transform our entire approach to our global grand challenges and enable the full engagements of each country, individual, industry and stakeholder group that has a desire to participate.

What gets me *really* excited in approaching sustainability from this angle, is that we're not only paving the way for new solutions to have maximum impact, but it can also be used to reinvigorate existing technologies, products, services and infrastructure.

And I believe these planet centered outcomes are only possible because of the emergence and subsequent convergence of the Five Transformational Forces of Digital Sustainability. These forces of transformation are now bridging the divide between our cognitive, social, economic, governance and natural systems. We finally have the potential to catalyze a true paradigm shift, if these forces can be harnessed in time.

Choosing the right path

So, we have the digital tools thanks to the Five Forces of Digital Transformation. And we know they can create transformational impact – we've been through many examples together so you could see or rather, read about them in action. And we've got a firm grasp of how different things can be for us if we are to utilize these tools to create a planet centered operating system.

Now to apply them to infuse humanity's better qualities and all the necessary sustainability values capable of the kind of transformation that will secure planetary health and the future of our species... but how?

Good question. I wondered when that one might pop up.

Unfortunately, the answer is by no means as simple as applying a cookie cutter solution to each use case. Every technology has hundreds of different use cases that confront us with unique ethical dilemmas, value choices and risks based on specific user needs.

Many elements of our technological toolkit are applied to solving specific problems locally – while others have the potential to scale globally. Therefore, we need to think about designing a process where we can make these decision-points as clear and transparent as possible as well as consider the consequences of global application once exponential adoption kicks in. We also need to think carefully about how we make the transition from business as usual to 'planet centered' and the consequences that might occur if both use cases for the Five Forces of Digital Transformation continue to co-exist.

We also need to be aware that our use of digital technology will fundamentally affect the way that we experience a specific issue and become emotionally connected to it. Technology may separate us from our own humanity and mediate our experience of reality in a way that simplifies everything into a binary set of 1s and 0s. However, many aspects of our humanity occur on a continuum, involving ambiguity and paradox as well as evolving perspectives and cascading social relationships. [5] We must ensure that the Five Forces of Digital Transformation remain tools that we use to manage problems in our physical reality – rather than shape and dictate that reality through a digital lens.

There are many action points to be addressed. And we will.

But the first key action point that needs our attention can be summed up in one simple word: awareness. In the context of avoiding the risks and unintended consequences of the dark side of these forces, awareness is king.

In the context of this book, awareness is all about understanding the ways in which sustainability values and human experience can be infused and/or imposed upon by business, governments, technology, products and services. This allows you an opportunity to gain further understanding as to how embedding these values, or omitting them completely, could be used to bring out the best but also potentially the worst outcomes when applied at scale.

In the majority of circumstances, multi-system impact is only built into the design of a solution as a function of standard business procedure, in order to ensure sound execution. What I am instead proposing is a major shift, in terms of using the Five Forces of Digital Transformation and their associated digital technologies to intentionally achieve planet centered whole-system transformation and impact, in a manner that advances civilizational sustainability.

If we are to embed the best aspects of humanity and the necessary values of sustainability into our technology, governments, products, businesses and solutions, we must make a conscious choice to address five foundational elements into each aspect of their design, build and diffusion.

That is why we need to move toward adopting a 'Planet Centric Design' methodology.

In this context, I define 'Planet Centric Design' as the framework to ideate, design, build and deploy anything (product, service, technology, process, policy) with a focus on positive, planet centered, 'whole system' outcomes at their core.

This proposed concept of Planet Centric Design is based on these five core principles:

1. Having a Planet Centered Mindset
2. Having a Planet Centered Mission
3. Setting Measurable Targets
4. Mitigating Risks and Impact
5. Magnifying Public Good Outcomes

Let's unpack each one in a little more detail together.

Having a planet centered mindset

Embracing sustainability and net-zero carbon goals starts with a mindset shift in three directions: (i) moving from a focus on shareholder value to stakeholder value; (ii) moving from a competitive approach into a more collaborative ecosystem approach and (iii) moving from a profit-driven orientation to a mission-driven orientation. Having a 'planet centered mindset' is the first core principle in applying the Planet Centric Design method and is driven by a new understanding that every actor has a key role to play, and that success will only be achieved through collaboration and systemic transformation. Cultivating this mindset will be easier than you think. Look for mentors or leaders in your industry who are crafting sustainable solutions and identify how these solutions/processes can be applied to your context, business model and value proposition. Be prepared to cooperate, learn from and share knowledge and resources with your peers, business community and core stakeholders in terms of both successes and failures – every case study is valuable in this collective journey. Work with other organizations in your sector to build an enabling environment so that sustainability solutions have a level regulatory playing field and can achieve global scale. Find digital solutions to sustainability that can accelerate actions at speed and scale.

Having a planet centered mission

Being clear on your mission is the second core principle for applying the Planet Centric Design method. As organizations adopt sustainability as a core value, they need to adapt their mission statements to reflect this new orientation and the need to balance profit, people and planet. The new planet centered mission should be aligned to the core competencies, assets as well as the 'superpowers' of an organization and will need to be reflected in the business model. This new mission must be adopted by executive leadership and cascade down to all business units. Sustainability is a journey involving multiple incremental steps and every part of an organization. It involves both mitigating risks to an organization as well as capitalizing on opportunities for new markets, products and services. The key is simply starting with a bold mission and making public commitments about core sustainability values and targets that you know you can make viable progress toward.

Setting measurable targets

Ensuring your progress toward your mission can be measured is the third core principle to apply when using the Planet Centric Design method. Once an organization has adopted a new mission with sustainability at the core, the first key step forward is to adopt goals and targets that are feasible, measurable and transparent. It is useful to anchor these goals in the language of the Sustainable Development Goals and the Paris Agreement as these taxonomies are increasingly used across the public and private sectors serving as a common language and measurement framework. Transparency is the single most important 'secret sauce' here because it will help organizations measure the direction of travel as well as compare the costs and benefits. Make sure there is coherence between your targets and your business model so that they don't cannibalize each other. Environment, Social and Governance (ESG) principles and frameworks are a good starting point – but they don't always go far enough in terms of capturing bold sustainability opportunities that combine profit, people and planet. Consider becoming B-CORP or Earthmark certified to help focus on the public good that can be achieved. Once organizations identify how they will measure risks, opportunities and impacts they will need to report on progress to shareholders, stakeholders and the public as well as within upcoming digital disclosure processes and performance labels where applicable.

Mitigating risks and impacts

Mitigating risks and impacts is the fourth core principle when applying the Planet Centric Design method. Sustainability targets should always have two dimensions. On the one hand, organizations must begin to mitigate environment and climate risks and impacts to their business operations. This includes a commitment to full cost accounting and internalizing any environmental externalities on their balance sheets. Organizations should be able to increasingly demonstrate compliance with ESG frameworks and progressively demonstrate progress in addressing key risks. Organizations like Project Drawdown are dedicated to helping businesses achieve net-zero emissions targets as quickly, safely, and equitably as possible.

Magnifying public good outcomes

And finally, the fifth core principle to consider when applying the Planet Centric Design method relates to having a commitment to magnify public good outcomes. When you have implemented a solution that is making an impact, don't just share it with your peers and members of the business community, actively seek out ways to join forces with the global community actively working on similar outcomes. Work with like-minded organizations to accelerate the impact being made so it has even farther scope, reach and scale. By working collaboratively with other leaders, organizations can forge the sustainability vision and increase the ambition level of an entire sector. Join coalitions of the willing that have a shared vision and participate in global grand challenges with measurable short-term commitments and public reporting frameworks. Keep an eye out for the Coalition for Digital Environmental Sustainability (CODES) and their efforts to accelerate a digital planet for sustainability by integrating sustainability values, data and metrics into the very codes of the digital economy.

Addressing the human-in-the-loop factor

Like it or not, human values are embedded in our digital technologies and the algorithms that drive them. In some cases, it is intentionally built into the technology, and in others; it happens via osmosis. Consider the values expressed by a Tesla versus a petrol powered SUV. You might argue that they are both just cars. But what of the values that were embedded in them during their design and build? One technology speaks to 'we are trying hard to adopt sustainable mobility' and the other says 'the environment did not factor into the design and build of our vehicle'.

By default, the technologies we create from harnessing the Five Forces of Digital Transformation and/or how it is used become an extension of the conscious and subconscious values of the funder, coder, designer, decision-maker or the end user – or all five. Given this, it is reasonable to suggest that any new technology needs to be focused on generating ethical considerations.

Whose values will be reflected and through which cultural lens? What makes these values more important than others? How do we reflect them in the development of our technology? How will embedding these values impact on other people and our planet? How can we disclose, validate and audit these value claims? How do we deal with the fact that different use cases for our technologies might have specific ethical dilemmas? How do we combine global values with local ones?

These are not small questions and to find the answers we, as a human race, will need to universally align on some pretty big ideas. Though we can no longer simply ignore the need to ensure our technology is built in ways that can uphold a values system that drives healthy outcomes for all of civilization and our planet.

Luckily, we don't have to start from scratch. Many of the international agreements brokered by the United Nations already reflect a minimum set of shared principles and values. A universal baseline. These cover human rights, climate change and sustainable development among many other topics. Nations of the world have already agreed that these aspirations should form the basis for global cooperation and be the outcomes that we strive to achieve from the economy.

Many of these agreements are not perfect but they are a universal starting point that have a measure of legitimacy and are broadly accepted as important and necessary in any future civilization. They are something to build on. A floor rather than a ceiling.

Another consideration is all about inclusion and equity. Digital technologies are rapidly transforming society, simultaneously allowing for unprecedented advances in the human condition whilst also giving rise to complex new social challenges and digital divides. However, they will never contribute to global sustainability if they merely magnify existing inequalities or exclude certain groups from accessing their potential benefits.

Currently more than half the world's population either does not have access to the internet or can afford to use only a fraction of its potential. And for those that are connected, the digital economy doesn't always reflect an inclusive approach or secure a fair distribution of benefits as part of the design. Whatever technology we build into the future needs to disrupt this problem to ensure the values of inclusion and equity become operational in its design.

Avoiding bias in the design of your technology often goes hand in hand with the question of equity. Humans have more than 180 cognitive biases built into our own operating system. These can be hardwired into our technology solutions without our awareness and can lead to some significant consequences.

The uproar over facial recognition is a perfect example. Researchers have found that leading facial recognition algorithms have different accuracy rates for different demographic groups, gender and race. So how can it be safely relied upon? As it stands today, it can't, but so far that hasn't stopped us from using it anyway. But what happens when technologies that have embedded biases are used to create new technologies as part of a technology stack? How do we avoid the core DNA our technology is built from becoming irreversibly corrupted with our own imperfections?

This is exactly what IBM's AI Fairness Toolkit aims to support technology focused organizations with. It does this by helping developers test for algorithmic fairness and bias on an ongoing basis as well as implement corrective measures. Checklists are also emerging as an avenue to help stakeholders identify and manage this risk. The Ethical OS Toolkit uses a structured question set to help kick-start the conversation around identifying potential risks and social harms from algorithmic bias.

Finally, we need to consider that digital technologies do not fully capture everything in the world that needs to be considered to advance sustainability. As Albert Einstein observed *"Not everything that can be counted counts, and not everything that counts can be counted"*. Humans experience the world in many diverse ways – these are reflected in our perceptions, cultural norms and shared stories. Technology can't deal well with complex concepts like love, fear and jealousy but these factors need to be accounted for. And how about more spiritual ways of experiencing life – such as those of indigenous groups around the world. These pose huge challenges to communicate let alone digitize but need to be considered and included in the equation nonetheless. [6]

These transformational forces are game-changing in their ability to fundamentally re-engineer and interconnect our operating systems with our natural operating system, in a manner that can underpin planetary sustainability. Once we shift into this broader planet centered mindset and whole-system lens, I think it will be impossible to "see" the world in any other way.

References

1. Rimanoczy, I. (2021). *The Sustainability Mindset Principles: A Guide to Developing a Mindset for a Better World*. Routledge.
2. Rushkoff, D. (2019). *Team Human*. W. W. Norton & Company.
3. SamHarris.org. (2020). *Sam Harris Podcast #218 – Welcome to the Cult Factory*. Available at www.samharris.org
4. United Nations Environment Programme. (2019). *Measuring Progress: Towards Achieving the Environmental Dimension of the SDGS*. Available at www.unep.org
5. Rushkoff, D. (2019). *Team Human*. W. W. Norton & Company.
6. Yunkaporta, T. (2020). *Sand Talk: How Indigenous Thinking Can Save the World*. Text Publishing.

11 May the transformational forces be with you

Peering over the edge of possibility

It is my belief that the next phase of our journey of life on Earth will require the rapid evolution of our collective intelligence to catalyze the exponential transformation of our operating systems. A feat I believe is only possible if we are to harness the power of Five Forces of Digital Transformation and make the choices necessary to consciously shape them in a direction that remains wholly focused on outcomes that are planet centered.

Unlike every other phase of our human evolution, these next changes that will shape how we evolve will not happen by chance or because of the discovery and initiation of one person or a single organization. Nor will these changes have the luxury of time to take root over a long period; like we could rely on back in the era of linear progression. No, our next evolutionary steps as a species need to happen rapidly, deliberately and will be made with conscious specificity, driven intentionally by us, the coalition of the willing.

This is a massively daunting task, yes, but also extremely exciting because the world is finally waking up, recognizing the severe impact we have made on the planet and the urgency in which we need to act in order to have a hope of reversing the damage we've caused.

The coalitions of the willing are emerging with a shared resolve that the transitions needed will be largely catalyzed by the Five Forces of Digital Transformation. They are prepared to take on the responsibility to devise the strategies and solutions needed to set things right, and to put the action steps in place to execute on this.

It is time for us to discover how to foster and scale these five forces. How to supercharge them using our collective intelligence to further unlock the innovative power digital technology wields, so that the solutions we implement can begin to move at the speed and scale of the global problems we face and enable our operating systems to keep pace with it all.

So, what needs to change? How can we now leverage the Five Forces of Digital Transformation and the onslaught of new technologies to achieve an equitable and sustainable human civilization in the next ten years and beyond? What actions must be taken in the next 12–18 months to enable this

DOI: 10.4324/9781003187523-14

transition? And how do we straddle the pace at which the Earth needs us to change with that of the rate of how people's behavior can change, based on our historical experiences?

The time has come for humanity to tap into the global collective on a much larger scale so that together we can tip the scale and hit critical mass in order to see governments, industry, the private sector, citizen science and civil society join the global effort to make sustainability the most attractive and *only* strategy for business and life on Earth.

It probably sounds like I am suggesting transformational change on the largest scale ever attempted in human history across every sector of our economy, every layer of society, every way of being and doing and earning and living.

Yes. That's because I am.

Moving from overwhelm induced apathy toward action and outcome

As complicated and complex as achieving this shift to planet centered sustainability is, it is, in my view, completely possible.

However, oftentimes, when we are faced with transitional and transformational elements that are both complicated and complex; fear, overwhelm, apathy and possibly even panic can set in and freeze us in a state of procrastination and inaction. I would consider it to be an absolute tragedy if, after all we have unpacked together, I thought for a moment you would be left in that state upon the completion of reading this book.

So how can I ensure we part ways with you feeling inspired, motivated, well informed and excited about the road ahead and the role you can play in it?

Well, in my nearly 22 years of involvement in rapidly changing, global environments I can share that there are a few tried and tested methods I have put into practice whenever I have personally felt a state of overwhelm induced apathy kick in. Let's hope they can also make a difference for you, so we are able to remain in a state of forward momentum long after you turn the last page with me.

The first? Is to simply take action. No-one ever articulated that sentiment better than David J. Schwartz when he penned the words, "*Action cures fear*". [1] Yes, it is time to act. Now. Before we have all the answers, or even a firm grasp on exactly where this journey might take us. We know there isn't a roadmap and nor is there a compass, but that's just how the approach to solving wicked problems that have no working solution usually get started. No knowns, only unknowns.

We need to adopt the mentality of being a giant start-up; capable of being collaborative, flexible, self-organizing and agile and willing to embrace the ambiguity involved in that agility. Innovation can only thrive in an environment capable of such things and whilst we will need to be considerate of the risks that can be inherently involved in innovation, we also need to embrace its ambiguity, believe in the process and just get started; focusing first on the more simple aspects that can catalyze as big an impact as possible and then steadily working our way toward the more complex and difficult to execute upon. Our planet and people are depending on it.

Our future world needs to be one built on a foundation of sustainability and equity and, as we have established, our current architecture for civilization won't take us there. Our only option is to embrace innovation and be prepared to fail, again and again, learn fast from those failures, then fail even faster again, ensuring every lesson we learn is learned the first time. Then just like Thomas Edison did with the light bulb and the Wright brothers did with flight, believe that eventually, we will succeed.

The second thing we need in order to successfully approach the starting line is a new perspective.

I want to suggest that perspective needs to account for two overarching considerations:

1. That we need to begin with the end in mind. What does this successful transformation really look like? What does it ultimately signify for civilization? Do we truly understand the bigger picture trajectory this transformational journey puts us on? and
2. Even our most complex and complicated aspects of our transformation to a sustainable ecosystem can be instantly viewed as being far more viable, achievable and simple when put in the context of even larger, more complicated and complex goals.

The process of transition and ultimate transformation to a conscious and sustainable civilization will unlock our ability to act as a collective intelligence and enable us to organically evolve beyond solving the wicked problems of the climate crisis and into a planet centric approach and effort to address all categories of humanity, life, science and planetary stewardship. This will be underpinned by a range of new technologies that help agents of change find each other, build deep trust and take swift collective action.

Therefore the far greater vision, the ultimate moonshot, is found in imagining what a civilization based on our harnessing of these Five Forces of Digital Transformation would look like in 2100. What do you think, shall we put our exponential goggles on together, one last time?

By 2100, through this exponential lens, I can see we will have achieved a fully integrated, self-regulating, sustainable civilization. People and planet are symbiotic in their relationship, balance restored. Spaceship Earth has only green lights across its dashboard. The natural evolutionary by-product of the advancements that helped us get to that point would then enable us to transform from our planet centric focus to that of a universe centric focus and we shift into a new age. Perhaps it will be known as the Terracene; the age within which humanity was in balance with all planets and lifeforms. We begin advancing in the realms of intergalactic stewardship, interplanetary habitation, interdimensional travel and so much more. The ongoing evolution of the human race is certain and we are now entirely decoupled from negative environmental impact, in turn ensuring the survivability of our species and enabling us to unlock limitless potential in these transformational forces for where the human species could go next.

It is one hell of a moonshot, yes, but perspective shifting exercises like this allow us to look at and approach implementing transformation for a global effort to solve our global grand challenges from a view that has a much higher vantage point. One where we can see the otherwise giant leaps we need to make toward a sustainable civilization as palatable, much smaller, far simpler steps in a much, much longer journey of discovery and human potential.

Yes, it is so easy when faced with such an overwhelming set of complexities and considerations to feel like there is no way to effectively craft solutions. However, the secret sauce is in being willing to expand our minds and consciousness in order to be open to the possibility of being able to create these solutions. In fostering this open mindset – we then naturally fall into the process of beginning to take the necessary steps to find the very solutions we thought impossible.

The most important transformational force of them all

On that note dear readers, I want to thank you for being brave enough to keep your exponential goggles on, right up until the very last page of this book. We've explored so much together but now it is time to say goodbye. However, before we part ways, I would like to leave you with one final thought.

Please, above all else, remember that integrating everything you've learned about our operating systems, the Five Forces of Digital Transformation and how to ensure these forces converge for planet centric outcomes, along with the strategies mentioned in this chapter to keep you inspired, focused and action orientated, are all designed to do one thing. To catalyze the sixth and most important transformational force of them all:

You.

So it is with hope in my heart for all the ways in which you will turn these words that fill these pages into action based outcomes for our people and planet, that we part ways. Go well.

May the force(s) be with you.

Reference

1. Schwartz, D. (1987). *The Magic of Thinking Big*. Vermilion.

Index

For Product Safety Concerns and Information please contact our EU
representative GPSR@taylorandfrancis.com
Taylor & Francis Verlag GmbH, Kaufingerstraße 24, 80331 München, Germany

www.ingramcontent.com/pod-product-compliance
Lightning Source LLC
Chambersburg PA
CBHW070345270326
41926CB00017B/3995